NEUROSCIENCE METHODS

NEUROSCIENCE METHODS

A Guide for Advanced Students

Edited by

Rosemary Martin

Australian National University
Canberra, Australia

harwood academic publishers
Australia • Canada • China • France • Germany • India • Japan
Luxembourg • Malaysia • The Netherlands • Russia • Singapore
Switzerland • Thailand • United Kingdom

Published in The Netherlands by Harwood Academic Publishers.

Amsteldijk 166
1st Floor
1079 LH Amsterdam
The Netherlands

British Library Cataloguing in Publication Data

A catalogue record for this book is available from the British Library.

ISBN 90-5702-245-1 (softcover)

CONTENTS

PREFACE

Scientists choose to study the brain in many different ways. Some prefer to work at the molecular level, unravelling the contribution of variations in gene expression or intra-cellular proteins to normal function or neuropathologies. Others are engaged in understanding the electrical properties of neurons or the chemistry of the brain and still others prefer to study the overt behavior of an organism and surmise about the design of the 'black box' which produces such behavior.

Not surprisingly, a large number of very specialized techniques have been developed to allow these avenues of study. Most undergraduate, honours and new PhD students find the plethora of techniques bewildering. As a consequence they have trouble understanding original papers and find it difficult to critically appraise them.

To help solve this problem, I asked a number of colleagues to each write a few pages about a method they used to study the nervous system. I particularly requested that they directed their writing towards an outline of the methodological principles and potential pitfalls of the technique they chose to describe.

It is always difficult to know where to stop with a book such as this. In the main, the techniques described are those used in animal experiments but I have chosen to include a few chapters on electrophysiological measurements in humans. Descriptions of some modern imaging techniques such as NMR, PET and SPECT and behavioral methods will have to wait, perhaps for another edition.

This book is the outcome of the labors of my many colleagues and I thank them all for so generously giving their time. I also wish to thank Garry Rhoda for his excellent editing of diagrams and Audra Johnstone for preparing the manuscript in its final form.

CONTRIBUTORS

John M. Bekkers
Division of Neuroscience
John Curtin School of Medical Research
Australian National University
Canberra, Australian Capital Territory, Australia

John Clements
Division of Neuroscience
John Curtin School of Medical Research
Australian National University
Canberra, Australian Capital Territory, Australia

Paul Cooper
Division of Botany and Zoology
The Faculties
Australian National University
Canberra, Australian Capital Territory, Australia

Anna I. Cowan
Division of Neuroscience
John Curtin School of Medical Research
Australian National University
Canberra, Australian Capital Territory, Australia

Michael E. Crouch
Molecular Signalling
Division of Neuroscience
John Curtin School of Medical Research
Australian National University
Canberra, Australian Capital Territory, Australia

Greg M. de Plater
Division of Neuroscience and Division of Biochemistry and Molecular Biology
John Curtin School of Medical Research
Australian National University
Canberra, Australian Capital Territory, Australia

Fiona M. Freeman
Division of Biochemistry and Molecular Biology
John Curtin School of Medical Research
Australian National University
Canberra, Australian Capital Territory, Australia

Susan Howitt
Division of Biochemistry and Molecular Biology
The Faculties
Australian National University
Canberra, Australian Capital Territory, Australia

Sven Johanson
Division of Neuroscience
John Curtin School of Medical Research
Australian National University
Canberra, Australian Capital Territory, Australia

Pauline R. Junankar
Division of Biochemistry and Molecular Biology
The Faculties
Australian National University
Canberra, Australian Capital Territory, Australia

Brian Key
Laboratory of Molecular Neurodevelopment
Department of Anatomy and Cell Biology
University of Melbourne,
Parkville, Victoria, Australia

Joseph I. Kourie
Department of Chemistry
The Faculties
Australian National University
Canberra, Australian Capital Territory, Australia

Gary Lacey
Therapeutic Goods Administration
Commonwealth Department of Health and Family Services
Canberra, Australian Capital Territory, Australia

Derek Laver
Division of Neuroscience
John Curtin School of Medical Research
Australian National University
Canberra, Australian Capital Territory, Australia

Andrew J. Lawrence
Department of Pharmacology
Monash University
Clayton, Victoria, Australia

Arnie Leon
Division of Neuroscience
John Curtin School of Medical Research
Australian National University
Canberra, Australian Capital Territory, Australia

Vaughan G. Macefield
Prince of Wales Medical Research Institute
Randwick, New South Wales, Australia

Nishith Mahanty
Department of Human Physiology
Medical Sciences
University of Newcastle
Newcastle, New South Wales, Australia

Lauren R. Marotte
Developmental Neurobiology
Research School of Biological Sciences
Australian National University
Canberra, Australian Capital Territory, Australia

Rosemary L. Martin
Division of Botany and Zoology
The Faculties
Australian National University
Canberra, Australian Capital Territory, Australia

Peter G. Noakes
Department of Physiology and Pharmacology
University of Queensland
Brisbane, Queensland, Australia

Sharon Oleskevich
Visual Sciences
Research School of Biological Sciences
Australian National University
Canberra, Australian Capital Territory, Australia

Stephen P. Perrett
Department of Neurobiology
Civitan International Research Center
University of Alabama at Birmingham
Birmingham, Alabama, USA

Bill D. Phillips
Institute for Biomedical Research
Department of Physiology
University of Sydney
New South Wales, Australia

Paul M. Pilowsky
Department of Neurosurgery
Royal North Shore Hospital
St Leonards, New South Wales, Australia

Laura J. Reece
Developmental Neurobiology
Research School of Biological Sciences
Australian National University
Canberra, Australian Capital Territory, Australia

Peter R. Schofield
Garvan Institute of Medical Research
Darlinghurst, New South Wales, Australia

Christian Stricker
Institute of Neuroinformatics ETH/Uni
Gloriastrasse 32
CH-006, Zürich, Switzerland

Maria Vidovic
Developmental Neurobiology
Research School of Biological Sciences
Australian National University
Canberra, Australian Capital Territory, Australia

Trichur R. Vidyasagar
Centre for Visual Science and Division of Neuroscience
John Curtin School of Medical Research
Australian National University
Canberra, Australian Capital Territory, Australia

SECTION 1

IN VITRO PREPARATIONS

CHAPTER 1

SLICES OF BRAIN TISSUE

C. Stricker
Institute of Neuroinfomatics ETH/Uni, Switzerland

INTRODUCTION

The brain slice technique was first applied to metabolic studies of small tissue sections as early as 1920. With the advent of the microelectrode technique at the beginning of the 1950s it was possible to characterize the resting membrane potentials of cells contained in freshly sliced tissue and compare them to brain cells *in situ*. It was soon realized that such slices provided new routes for the study of synaptic phenomena. Towards the end of the 1960s the use of surviving, metabolically maintained tissues from the brain for electrophysiological and pharmacological studies was becoming an accepted, valued and widely applied technique. But it was not until the advent of the whole-cell recording technique in the mid 1980s and its application to sections of brain tissue that the brain slice technique became very popular.

Brain slices weighing some 10 mg can provide access to fine structures and thus allows functional analysis of parts of the brain. Isolates of this size comprise some 10^4 to 10^5 cells. A large cerebral neuron may synapse with 10^3 to 10^5 other cells. Thus the cell pool provided in a brain slice may be the minimum unit size needed for adequate display of the connectivity of a cerebral neuron in its adult environment. Despite the trauma of their formation, adequately prepared slices approach the status of biological entities.

However, it is unrealistic to consider sliced tissue to be 'normal', no matter how skilfully and carefully the slices have been prepared. Slices are isolated tissue, without normal inputs, immersed in an artificial environment. Investigators must weigh the potential advantages with the obvious disadvantages as they apply to their particular problem. Advantages of the slice are: visibility, technical accessibility, stability and ease of use. Disadvantages are: loss of normal input pathways, shearing of dendritic processes and axons, tissue debris around the cutting surface mixed with healthy cells, slow release of cellular enzymes and ions from damaged cells, and altered metabolic state.

Depending on the experimental needs several variants of the technique have been developed and are being used. One distinction between the variants relates to how thick the slices have been cut. The thin slice technique was developed to allow visualization of individual cells in slices of less than 250 μm while the thick slice technique is used in experiments where connectivity and maintenance of normal dendritic structure are crucial for the study. Once prepared, slices can be kept alive in various media for hours or weeks. Ultimately slices can be kept in culture.

In the following paragraphs the slicing procedure and the knowledge surrounding it will be explained, with a bias towards the hippocampal slice. This bias results from the fact that the hippocampal slice is the most commonly used slice preparation today (Figure 1.1). The attraction of this slice is due to its clearly layed-out cytoarchitecture, where the cell bodies lie in various clearly visible cell bands, and dendrites make contact with fibers from known origin. A lot is known about the histology, as well as the pharmacology of the different areas of the hippocampus. Even though the hippocampus is the most widely used slice preparation, many others have been established in the last ten years. It is theoretically possible to cut any sort of slice from any region of the central nervous system.

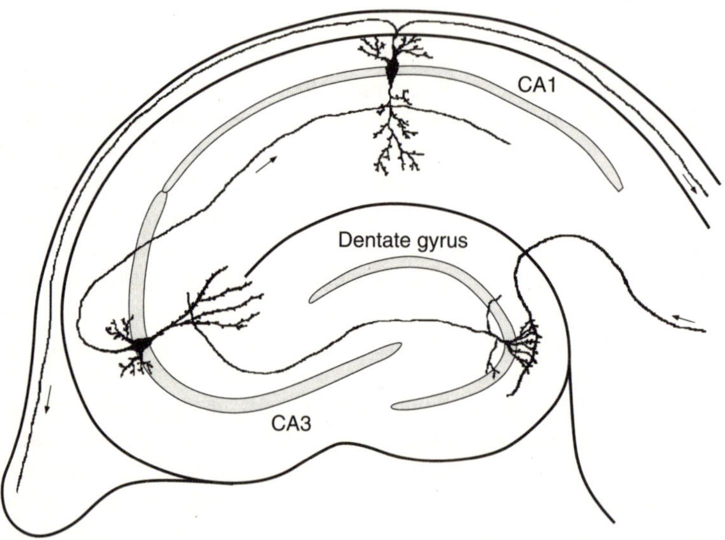

Figure 1.1 Schematic drawing of a parasagittal hippocampal slice. Shaded areas indicate where the cell bodies of the principal excitatory cells are found, i.e. in areas CA1, CA3 and the granule cell layer of the dentate gyrus. The remaining area within the slice contains interneurons and axons making synaptic contact onto the dendrites of the pyramidal cells. The flow of excitation within the hippocampus is indicated by the small arrows.

SLICING PROCEDURE

'It is perhaps on this topic more than others where myths, unfounded dogmas, and notions based on intuition or anecdotal evidence tend to influence the choice of method' (Alger et al., 1984). Despite the many different procedures employed, the main goal is to prepare a slice of tissue where the neurons, fibers, synapses and glia that are important to the experiment are in a viable condition.

The animals used in preparing slices are most often small rodents: guinea pig, rat and mouse. It appears that younger animals produce better results than older animals mainly because they are more resistant to the traumatic and ischaemic insult of the slicing procedure. It seems that the speed of dissection is not nearly as important as the care taken in removing and slicing the tissue. Several studies point to the fact that the actual cutting of the tissue is the critical step. The removal of the brain tissue is done after decapitation of the animal or during deep anaesthesia. Decapitation tends to be slower but less bloody, and it seems to yield superior results.

Once the tissue is removed, it is normally cooled down to temperatures around 2–4°C by placing it in ice-cold oxygenated artificial cerebrospinal fluid (ACSF) to minimize metabolic activity. The piece of tissue is then cut with a scalpel to obtain the desired tissue orientation and then glued onto a stage using cyano-acrylic glue. Cutting is then done in ice-cold ACSF using a vibratome, a mechanical instrument which cuts by slowly moving a laterally vibrating blade through the brain tissue. These instruments were originally designed for the preparation of histological specimens. The rate of advance and vibration amplitude of the blade are best set at the maximum values that will permit rapid cutting without compressing or 'pushing' the tissue. Small blocks of agar (2–5% made up in ACSF) can be glued onto the stage for additional support.

A 'standard' slice is cut at 400 μm thickness. This thickness is a compromise between retaining the cytoarchitecture and visibility, and the diffusion distances for oxygen and glucose. It can be shown that the limiting thickness of a cerebellar slice is about 450 μm at 37°C. Regions of the slices that are thicker than this value exhibit centrally-located necrotic cells, suggestive of hypoxic damage. This limiting thickness may vary in different brain regions according to the particular tissue demands.

After cutting, the slices usually need to be trimmed away from the surrounding tissue with fine scissors and forceps. The slices are then incubated at a temperature of around 36°C for at least forty minutes. Oxygenation and normal pH are maintained by bubbling the ACSF with 95% O_2/5% CO_2. This allows the tissue to 'recover' from the damage imposed by the preparation and adjust to the new extracellular milieu as well as to the changed metabolic activity. It has been suggested that during the incubation period cellular enzymes are released which help 'soften' the surface of the slice. This seems to be important for whole-cell recording. Following the recovery period, the slices are maintained at room temperature to keep metabolic activity low.

SLICE CHAMBERS

For experimental use slices must be kept in an environment providing appropriate oxygenation, pH, osmolarity, and temperature. In addition, depending on the techniques used, it is necessary to have excellent visual control, good mechanical access and stability. Most commonly used chambers allow the superfusion of ACSF across the slice. This imposes special demands on the mechanical stability of the superfusion system.

There are two different superfusion chamber designs where the slice either rests on a net at the gas–liquid interface (so-called 'interface chamber') or is totally submerged ('submersion chamber'). The best design depends on the particular experimental requirements. Submersion chambers are normally used for whole-cell patch recording, whereas interface chambers are better suited for monitoring extracellular fields. It is known that field EPSPs are bigger in interface chambers probably due to the fact that in a submersion chamber the current can flow more easily into a bigger superfusion volume than if its extracellular space is restricted as it is in the interface chamber. Most of the slice chambers are temperature controlled and the superfusion rate of the ACSF is at least 1 ml/min. The flow rate determines the O_2 and CO_2 escape by diffusion as the perfusate travels along the tubing supplying the chamber. Flexible tubing is normally made out of Tygon™ which has low diffusion constants for O_2 and CO_2.

Mechanical stability of recordings in slices is determined, among other things, by the inflow and aspiration rates of ACSF and the balance achieved between them. Drainage is the most difficult factor to control. Mechanical stability can be disrupted by occasional gas bubbles which can be trapped in additional reservoirs not directly connected to the superfusion volume in the chamber. For recordings of any sort, it is important to keep the slice fixed to the bottom of the chamber. This is achieved by laying a grid of parallel nylon threads glued on a U-shaped flattened platinum wire on top of the slice (Edwards et al., 1989).

ACSF

Comparisons of the ionic composition of ACSF with *in situ* solution shows that the most significant differences are in the K^+, Ca^{2+} and Mg^{2+} concentrations. The higher concentrations used *in vitro* may well affect physiological properties of the slice. It is well known that divalent cations like Ca^{2+} and Mg^{2+} exert a stabilising effect on excitable membranes which will raise action potential thresholds. Further, both Ca^{2+} and Mg^{2+} have a profound effect on synaptic transmission. Note that the effective Ca^{2+} and Mg^{2+} concentration within the slice is smaller than the concentration in the ACSF due to the fact that a portion of Ca^{2+} and Mg^{2+} is chelated with phosphates in the solution as well as bound to proteins within the slice.

A normal mammalian ACSF (rat) contains about (in mM): NaCl 124; KCl 3; $NaHCO_3$ 26; NaH_2PO_4 2.5; $CaCl_2$ 2.5; $MgSO_4$ 1.3, glucose 10.6. Different laboratories use slight deviations of this recipe. The important point is that the osmolarity of the solution is between 280 and 320 mOsm/l. Glucose is the primary energy substrate in this ACSF. pH is adjusted between 7.2 and 7.4. The final pH value is obtained by bubbling with CO_2 (bicarbonate/CO_2-based buffering system). Bubbling the ACSF is normally done with carbogen, a gas mixture of 95% O_2/5% CO_2.

For preparations like the adult guinea pig brain, it is essential to cut slices in ACSF where sodium has been replaced iso-osmotically with sucrose. The reasons for this are currently unclear but lowering of the transmembrane sodium gradient will influence many membrane transporters and will change overall excitability.

The above mentioned composition of ACSF reflects the basic requirements for maintaining healthy slices. Depending on the experimental needs, the composition might have to be adjusted considerably, and perhaps additional constituents and/or pharmacological agents added. For example, working on NMDA ligand gated channel usually requires the addition of the co-agonist glycine to the superfusion solution.

TEMPERATURE

Although the body temperature of small rodents is around 38°C, most investigators maintain the slices at 30–35°C in the experimental chamber. There are two reasons for this. Firstly, it has been found that preparations survive longer, and in a healthier state at the lower temperature. Secondly, the higher humidity resulting from warmer solutions leads to the formation of droplets on recording and stimulating electrodes. These tend to fall off and result in mechanical instability during recordings. Many people tend to work at room temperature. However, as most of the biological processes have a Q_{10} of about 2, working at room temperature will slow biological processes down to half their normal values. Most affected are ionic currents and synaptic transmission.

CELL VISUALIZATION TECHNIQUES

The real advantage of the slice is its accessibility, especially the visibility of structures such as cell body layers. If individual cells are being sought, slices are normally cut to a thickness of less than 250 μm (thin slice technique). Thin slices allow a greater optical resolution due to the smaller effect from light scattering. The slices may also be obtained from younger animals because they have thinner myelin sheets which also improves visibility. Slices are placed in a chamber on the stage of an upright microscope. Inverted microscopes are not preferred due to the problems encountered when trying to visualize the approach of recording electrodes and stimulating electrodes to the tissue. The cells and parts of the dendrites can be visualized using a 40 times high numerical aperture, long working distance water immersion objective. To further improve the contrast between different cells, Nomarski or Hoffman optical arrangements are preferred.

Newer approaches make use of the properties of infrared light (750–850 nm). The basic theory behind this technique is the fact that brain tissue is much more translucent to longer wavelengths. This increases the transparency for cellular visualization. Infrared microscopy (Dodt & Zieglgänsberger, 1990) can easily be implemented on a normal microscope by using a band-pass filter blocking light with a wavelength of less than 750 nm and longer than 1050 nm as the longer wavelength will heat the slice. The image produced by the optics is then projected to an infrared sensitive video camera and the image generated is displayed on a video monitor. Optical resolution can be further improved by digital image processing.

ASSESSMENT OF SLICE PREPARATIONS

'Despite worries about the suitability of slices for certain studies and the problems that an isolated *in vitro* preparation introduces into interpretation of data, investigators should remember that virtually all experimental preparations *in vivo* as well as *in vitro*, chronic as well as acute introduce interpretational difficulties. The most satisfying validation of slice phenomena has been the general finding that *in vitro* studies are similar to *in vivo* investigations. Clearly, this reasoning is somewhat circular, and our criteria for useful data are invariably arbitrarily set.' (Alger et al., 1984). There are, however, some helpful techniques to evaluate the suitability and viability of slices.

Histology

A clearly, well-defined cytoarchitecture, like that found in the hippocampus or the cerebellum can help in assessing slices. Without staining, highly translucent cell bands, ill-formed borders and 'mushy' consistency are indicative of swollen cells and are a bad omen. Obviously, there is always some necrosis present in a slice due to the cutting damage to the tissue. However, the necrotic tissue should not extend more than about 100 μm into the slice. Vital dyes, like trypan blue, are not taken up by viable cells, however they deeply stain necrotic tissue. Such dyes can give reliable guidance in the assessment of the viability of slices during an experiment. Post-experimental processing of tissue material based on histological stainings can help in assessing the extent of tissue damage, necrosis and oedema.

Electrophysiology

The assessment of electrical parameters of slices depends very much on the characteristics of the particular cells within the tissue and varies from brain region to brain region. Crude indicators include resting membrane potential, input resistance, amplitude of the action potential, and recording stability. More sensitive measures include the ability of the cells to produce a regular, rhythmic train of action potentials after the injection of a relatively small current. Damaged neurons will often respond with a single action potential at the onset of the current pulse.

Besides direct cellular parameters, amplitudes of extracellular fields reflect the synaptic action and are convenient for assessment of the overall state of a slice, or at least of small regions within a slice. However, there is no direct measure available since the size of a field potential is influenced by the resistance of the recording electrode as well as by the size of the tissue chamber and the depth of immersion. Large field potentials can also reflect cellular pathology; i.e. broad action potentials may reflect low temperature as well as hyperexcitability (high K^+).

Metabolism

These parameters are difficult to measure under experimental conditions and, therefore, will be of limited value in assessing the slice. However, it has been recognized that there are considerable differences between a slice and the *in situ* brain. The basis of these differences are unknown. A viable slice has been found to have ATP levels and O_2 consumption at 50–67% of its *in situ* value. Intracellular pH is about 0.3–0.4 units more alkaline and the intracellular K^+ to Na^+ ratio is 75% of normal (Alger et al., 1984). These values seem not to stem from the anoxic insult during preparation but may directly reflect the trauma of isolation.

LONG-TERM USE OF SLICES

Slices are normally used for about 12 hours after preparation and are considered to be best around 3–4 hours after slicing. Recently, however, the use of molecular techniques has made it imperative to keep slices for up to several days without the need for direct culturing techniques (for review see Lo et al., 1994). In such experiments, viral vectors are used to transfect certain cells in the slice with DNA or mRNA for subsequent protein expression (Pettit et al., 1994). The addition of balanced salt solution (see also culture technique), conditioned medium as well as certain amino acids to the incubation solution can keep the slices alive for several days.

By previous surgical, radiation, or chemical treatment of the slice, modified nervous systems can be produced. For example, modification after sectioning using microsurgical or optical techniques for removal of certain cell types can produce a piece of neural tissue suitable to answer very specific questions.

The combination of slices with the tissue culture technique has resulted in the development of organotypic cultures. This technique has been made available for different brain regions by Gähwiler (1981) who standardized the method in several aspects, including embedding material and culturing media. Recently, Stoppini (1991) has described a variant of the method which does not rely on embedding at all. The technique represents a hybrid between the explanted culture technique and the slice with its preformed connectivity. The cells survive for up to several months. Particularly interesting is an investigation of developing synaptic connections. It has been shown, for instance, that intrinsic GABAergic interneurons develop during the first weeks *in vitro* and form functional inhibitory synapses in the perisomatic region of pyramidal and granule cells (Streit et al., 1989). As such cultures form virtual monolayers they are very well suited for optical recording where the thickness of the normal slice might have restricted the optical resolution.

Additionally, tissue from other areas in the brain or the periphery can be cocultured with the slice. Such cultures have been described where spinal cord tissue is cocultured with dorsal root ganglion cells and embryonic muscle cells. It has been shown that the survival of the motoneurons in the spinal cord slice depends critically on the presence of the muscle cells (Spenger et al., 1991). Such slice cultures open new avenues to study development and the interaction between the neural tissue and its targets.

FUTURE DEVELOPMENTS

The unsurpassed success of the slice in studying the nervous system will lead to even broader applications. Having a more profound understanding of the metabolic requirements will eventually lead to ACSF that more closely mimics the neuronal environment *in vivo*. Such research will allow the maintance of viable tissue for longer periods. The use of molecular techniques will eventually lead to the design of transgenic slices for which the paper by Pettit et al. (1994) is an excellent example. Current transgenic tissue is normally obtained from animals which have been designed at early embryonic stages. The problem with such animals is that during development many mechanisms may very well compensate for changes imposed by the newly created genetic code. Slices could offer a real advantage since transfection could occur at various developmental stages and thus effects could be studied in the absence of any compensatory mechanisms. We will see new variants of this basic technique emerging in the near future.

REFERENCES AND FURTHER READING

Alger, B. E., Dhanjal, S. S., Dingledine, R., Garthwaite, J., Henderson, G., King, G. L., Lipton, P., North, A., Schwartzkroin, P. A., Sears, T. A., Segal, M., Whittingham, T. S. & Williams, J. (1984) Brain slice methods. In *Brain Slices*, edited by R. Dingledine, pp. 381–437. New York: Plenum Press.

Dingledine, R. (ed) (1984) *Brain Slices*. New York: Plenum Press.

Dodt, H. -U. and Zieglgänsberger, W. (1990) Visualizing unstained neurons in living brain slices by infrared DIC-videomicroscopy. *Brain Res.* **537**, 333–336.

Edwards, F. A., Konnerth, A., Sakmann, B. and Takahashi, T. (1989) A thin slice preparation for patch clamp recordings from neurons of the mammalian central nervous system. *Pflügers Arch.* **414**, 600–612.

Gähwiler, B. H. (1981) Organotypical monolayer cultures of nervous tissue. *J. Neurosci. Meth.* **4**, 329–342.

Kettenmann, H. and Grantyn, R. (eds) (1992) *Practical Electrophysiological Methods*. New York: Wiley-Liss Inc.

Lo, D. C., McAllister, A. K. and Katz, L. C. (1994) Neuronal transfection in brain slices using particle-mediated gene transfer. *Neuron* **13**, 1263–1268.

Pettit, D. L., Perlman, S. and Malinow, R. (1994) Potentiated transmission and prevention of further LTP by increased CaMKII activity in postsynaptic hippocampal slice neurons. *Science* **266**, 1881–1885.

Spenger, C., Braschler, U. F., Streit, J. and Lüscher, H. -R. (1991) An organotypic spinal cord—dorsal root ganglion—skeletal muscle coculture of embryonic rat. I. The morphological correlates of the spinal reflex arc. *Eur. J. Neurosci.* **3**, 1037–1053.

Stoppini, L., Buchs, P. -A. and Muller, D. (1991) A simple method for organotypic cultures of nervous tissue. *J. Neurosci. Meth.* **37**, 173–182.

Streit, P., Thompson, S. M., Gähwiler, B. H. (1989) Anatomical and physiological properties of GABAergic neurotransmission in organotypic slice cultures of rat hippocampus. *Eur. J. Neurosci.* **1**, 603–615.

CHAPTER 2

THE BRAIN IN A DISH: HOW TO STUDY ACUTELY DISSOCIATED AND CULTURED NEURONS

J. M. Bekkers
Division of Neuroscience, John Curtin School of Medical Research, Australian National University

INTRODUCTION

In order to study a structure as complicated as the nervous system, it is important to be able to tackle the problems at different levels of abstraction. Behavioral psychology is at one extreme of this scale of reductionism; the molecular biology of neuronal receptors is at the other. Between these extremes can be found large numbers of more or less idealized 'model systems' that are used by neuroscientists to simplify their experiments and make the interpretation of their results more tractable. This chapter will describe one class of such model systems: acutely dissociated and cultured cells from the mammalian nervous system. Three kinds of system will be outlined: (i) cultured neural slices, (ii) acutely dissociated neural tissue, and (iii) cultured, dissociated neural tissue.

The application of culture techniques to neurobiological questions has a long history, dating back to 1910, but it is only within the past twenty years or so that this approach has become part of the mainstream. One reason for this is the development of simpler, more effective protocols for keeping neurons alive in the dish for long periods. A major advantage of using cultured/dissociated neurons is the easy access that is possible when cells are freed from surrounding structures. A major disadvantage is that much of the architecture of the intact nervous system is lost. The pros and cons of the technique will be further discussed after the basics of each method are presented.

CULTURED NEURAL SLICES

Cultured neural slices are also known as organotypic slice cultures, because they retain much of the connectivity typical of the organ of origin. The approach is most often associated with Gähwiler and colleagues, who popularized it during the 1980s (Gähwiler, 1988).

The first step of the method is to prepare thin (200–700 μm thick) slices of the tissue of interest, using a razor blade mounted on some kind of chopping device. Of course, as for any cell culture, all preparation must be done using sterile technique, to avoid contamination with bacteria, yeast or mold. Sterile technique means all manipulations should be done in a laminar flow hood (which contains sterile, filtered air) and all solutions and instruments should be sterilized by heat, alcohol or filtration. A number of companies now provide sterile plasticware and solutions for culture use.

As a general principle, tissue from embryos or newborn animals is much easier to maintain in culture. This applies equally to dissociated cell preparations (see below). Possible reasons for this include the relative absence of connective tissue in young animals, permitting less damage during slicing or dissociation, and the relative resistance to hypoxia of juvenile tissue.

In the original Gähwiler technique, the slices are secured to glass coverslips by either collagen or a plasma clot (Gähwiler, 1988). These are then placed in sterile tubes, a small amount of culture medium added, and the tubes capped and put into a roller arrangement that slowly rotates the tubes at about 10 revolutions per hour in a 37°C incubator. The slices should be immersed in medium for one half of each rotation and out of it for the other half. The idea is that the tissue is kept both moist (through immersion) and oxygenated (through exposure to the air). A more recent variant of the organotypic technique places the slices on filter membranes which are then gently rocked in culture medium, without rotation (Romijn et al., 1988). It seems that by resting the slices on porous membrane, oxygenation is adequate without the need for air exposure.

Slices can survive in culture for weeks if they are fed with fresh medium once or twice per week. Over time, the slices may flatten out, until they are only several cells thick. Also, the structure in the original slices tends to become more diffuse as the neurons migrate and extend new processes. This inappropriate development is a possible disadvantage that will be discussed again later.

A word needs to be said about culture medium, the composition of which has a large bearing on whether or not the cultures will survive. Typically, medium for mammalian cultures comprises a bicarbonate-buffered high-sodium salt solution, essential amino acids and vitamins, glucose, antibiotics (in case one's sterile technique is not up to scratch), plus 10% fetal bovine serum (FBS) (Banker & Goslin, 1991). FBS is the 'magic' ingredient, in that it contributes all the trophic factors that seem to be important for cell survival but which are not well characterized. Unfortunately, FBS is both expensive and inclined to vary in effectiveness from batch to batch. Concern about variability has induced some researchers to develop 'serum-free' culture medium, to which various hormones, growth factors, etc. are added in place of the FBS. These attempts have met varying success; neurons have a reputation for being particularly fragile and fussy about their culture medium. Thus, FBS-containing medium is still the most commonly used, especially in research where developmental questions are not being specifically addressed.

ACUTELY DISSOCIATED NEURAL TISSUE

This approach is the simplest of the three techniques discussed here. Typically, pieces of tissue are cut from the region of interest, incubated in an enzyme (such as trypsin) to partially digest the connective tissue, and then gently agitated (e.g. by sucking it up and down a Pasteur pipette) to release single neurons. Of course, these neurons usually have their axons and finer dendrites torn off by the isolation procedure. However, principal dendrites often remain attached to the soma, and damaged processes quickly seal over, so cells remain viable for several hours if maintained in ordinary Ringer's solution. As mentioned above, the dissociation procedure is far less damaging for embryonic or newborn tissue, probably because of the relative absence of connective tissue in young animals. Adult tissue can be dissociated, but the yield of viable cells tends to be much lower.

Since neurons are freshly isolated and studied for only a few hours, sterile technique is not necessary, and nor is it necessary to deal with complicated culture media. On the other hand, it is obviously not possible to study synaptic connections with this system, because all connections are lost during the preparation. This technique is most often used for the study of intrinsic ionic currents, such as sodium or calcium currents (e.g. Sayer et al., 1993). An advantage of acutely dissociated neurons for this kind of work is that, because they lack many processes, the voltage clamping of the cell is likely to be more accurate (see Chapter 5). However, concern is sometimes expressed that these channels might be damaged by the enzyme used in the dissociation. Other pros and cons will be discussed later.

CULTURED, DISSOCIATED NEURAL TISSUE

This is the next logical step from acutely dissociated tissue. Following dissociation, individual neurons are resuspended in culture medium which is placed in culture dishes and maintained in a 37°C incubator. Of course, the dissociation and all other manipulations must now be done under sterile conditions. Within a few hours, the neurons settle onto the bottom of the culture dish and begin to put out new processes. Within a few days, functional synaptic contacts form between neurons, and cultures often show spontaneous electrical activity (Fletcher et al., 1991). Like organotypic cultures, dissociated cultures can be maintained for weeks by feeding them occasionally with fresh medium.

It should be noted that the method just described is also called primary cell culture, to distinguish it from the culture of neuronal cell lines. Cell lines are most often derived from tumors; that is, they are cells that have been transformed, or immortalized, so they continue to divide indefinitely. Examples of neuronal cell lines, available commercially, are neuroblastoma cells and pheochromocytoma (PC12) cells. Since such lines continue to proliferate, they must be periodically resuspended and diluted into new culture dishes to prevent overcrowding. For many kinds of experiments (e.g. those requiring large numbers of neurons, such as for neurochemistry or molecular biology) neuronal cell lines are very useful. However, they do not usually form functional synapses and there is always concern that a cancerous cell may tell us more about pathology than physiology.

In contrast to cell lines, neurons used for primary cultures are terminally differentiated, meaning that their phenotype has stabilized and they no longer undergo cell division. Although individual neurons become larger with time *in vitro*, they become less numerous as they die off. This sets a practical upper limit on the time for which these cultures can be maintained, usually a month or so. Primary cultures, like the brain, also contain glial cells. Unlike neurons, glia do continue to divide when placed in culture medium and, if allowed to do so indefinitely, will take over the culture. Hence, most researchers briefly treat their primary cultures with antimitotic drugs, like Ara-C (cytosine arabinoside) or FdU (fluorodeoxyuridine), to poison the replication apparatus of the glia. This seems to have no effect on the neurons. Organotypic cultures are also often treated with antimitotics for the same reason.

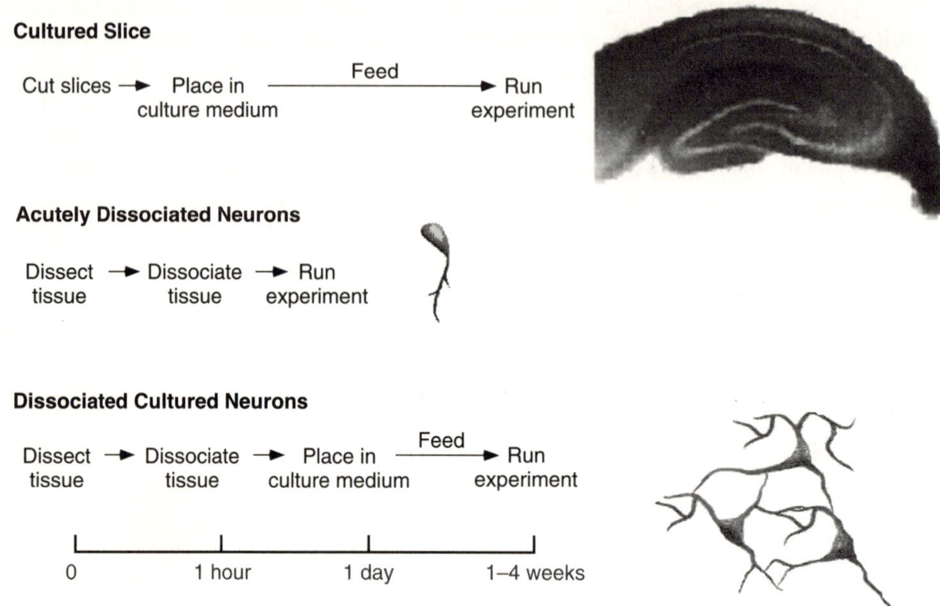

Figure 2.1 Summary of the methods outlined in this chapter.

ADVANTAGES AND DISADVANTAGES OF ACUTELY DISSOCIATED AND CULTURED NEURONS

As mentioned at the beginning, the *in vitro* systems described here represent a particular class of simplified models of the nervous system and, as for all models, caution must be exercised in extrapolating the model results back to the intact animal. Two kinds of simplification are made when neurons are dissociated or maintained in cell culture: first, all inputs from other parts of the brain are severed; second, growth occurs in a simplified, artificial environment.

The first of these simplifications is by no means unique to cultured preparations, since it also applies to acute brain slices (see Chapter 1) or other experiments in which afferent fibers are cut (e.g. in the spinal cord). The extent to which further disruption of connectivity occurs depends, of course, on the amount of dissociation employed. Organotypic cultures are the least disruptive in this respect, and acutely dissociated preparations are totally disruptive, although for the study of ion channels, say, this is probably irrelevant. However, cultures differ from acute slices in allowing a second level of simplification: in culture, further growth and development is allowed to occur in an artificial, defined environment. This fact has been the source of much of the criticism of culture as a useful model system. In the absence of trophic factors or other cues that are present in the intact animal (so the argument goes), development in cell culture may be inappropriate.

An example of the hazards of studying development in culture is provided by organotypic slice cultures. As mentioned earlier, neurons in these cultures tend to migrate out of their well-defined nuclei and extend new processes. Synaptic connections become far more profuse in these cultures, compared with acute slice preparations (Debanne et al., 1995). On the other hand, this observation can be turned into an advantage, by asking *why* synaptogenesis is more promiscuous in culture. Many important developmental studies have used cultures to address questions about what determines synapse specificity, how axons find their way, and, indeed, what determines whether a process becomes an axon rather than a dendrite (Dotti & Banker, 1987).

Whereas the number of synaptic connections in culture may be artifactual, the physiological behavior of each of those synapses seems to be unchanged from that of the corresponding synapses *in situ*. For example, the kinetics, pharmacology, ion selectivity and molecular make-up of post-synaptic glutamate and GABA receptors in culture are identical to those in acute brain slices (McBain & Mayer, 1994; Jones & Westbrook, 1995). These results indicate a class of questions, concerning synaptic transmission, that can be addressed in culture and reasonably extrapolated to the intact brain. However, caution must always be exercised. For example, there is evidence that an important class of glutamate-gated ion channel—the NMDA receptor/ion channel—may not be modulated in the same way in hippocampal cultures and acute slices (Bekkers et al., 1996).

CONCLUSION

To reiterate a point made at the beginning, a major advantage of using cultured/dissociated neurons is the easy access that is possible when cells are freed from the entanglements of surrounding tissue. The beauty of cultures is that a complex three-dimensional structure containing many billions of interconnected neurons—the nervous system—is reduced to a two-dimensional structure containing a much smaller number of cells. Pre- and post-synaptic neurons can be visually identified and recorded from, drugs can be locally and rapidly applied, the epileptic behavior of small networks of coupled neurons can be studied, and developmental changes can be followed with time-lapse videomicroscopy. Provided its limitations as a model system are recognized, cell culture offers a powerful approach to studying the operation of the nervous system.

REFERENCES AND FURTHER READING

Banker, G. and Goslin, K. (eds) (1991) Culturing Nerve Cells. In *Cellular and Molecular Neuroscience Series*, edited by C. F. Stevens. Massachusetts: MIT Press.

Bekkers, J. M., Vidovic, M. and Ymer, S. (1996) Differential effects of histamine on the N-methyl-D-aspartate channel in hippocampal slices and cultures. *Neurosci.* **72**, 669–677.

Debanne, D., Guerineau, N. C., Gahwiler, B. H. and Thompson, S. M. (1995) Physiology and pharmacology of unitary synaptic connections between pairs of cells in areas CA3 and CA1 of rat hippocampal slice cultures. *J. Neurophysiol.* **73**, 1282–1294.

Dotti, C. G. and Banker, G. A. (1987) Experimentally induced alteration in the polarity of developing neurons. *Nature* **330**, 254–256.

Fletcher, T. L., Cameron, P., De Camilli, P. and Banker, G. (1991) The distribution of synapsin I and synaptophysin in hippocampal neurons developing in culture. *J. Neurosci.* **11**, 1617–1626.

Gähwiler, B. H. (1988) Organotypic cultures of neural tissue. *Trends Neurosci.* **11**, 484–489.

Jones, M. V. and Westbrook, G. L. (1995) Desensitized states prolong GABAA channel responses to brief agonist pulses. *Neuron* **15**, 181–191.

McBain C. J. and Mayer, M. L. (1994) N-Methyl-D-Aspartic acid receptor structure and function. *Physiol. Rev.* **74**, 723–760.

Romijn, H. J., de Jong, B. M. and Ruijter, J. M. (1988) A procedure for culturing rat neocortex explants in a serum-free nutrient medium. *J. Neurosci. Meth.* **23**, 75–83.

Sayer, R. J., Brown, A. M., Schwindt, P. C. and Crill, W. E. (1993) Calcium currents in acutely isolated human neocortical neurons. *J. Neurophysiol.* **69**, 1596–1606.

CHAPTER 3

PAINTING OPTIMAL BILAYERS FOR RECORDING ION CHANNELS

J. I. Kourie
Department of Chemistry, The Faculties, Australian National University

INTRODUCTION

Optimal bilayers are artificial membranes that have negligible leak, high capacitance and stable electrical properties. The formation of lipid bilayer by the painting method, i.e. dispersing of a surface active lipid in a non-polar solvent, was first described by Mueller et al. (1962). It was used to identify the nature of ion transport pathways such as carriers or ion channels. The limitations (e.g. the presence of the solvent, time resolution and lack of control of vesicle fusion) and associated artifacts (arising from massive vesicle fusion, mitochondrial and bacteria contamination, and lipid oxidization) discouraged the wide use of this technique. However, after the advances made by using the patch-clamp technique (see Chapter 6), which established the presence of ion channels in biological membranes, it became apparent that the lipid bilayer technique is particularly useful in:

(a) controlling the composition and concentrations of chemicals in solutions on both sides of ion channels of internal membranes (e.g. sarcoplasmic reticulum (SR) and T-tubule membranes) which otherwise are inaccessible by patch-clamp electrodes;
(b) testing and probing the formation, structure and regulation of ion channels from purified proteins, synthesized peptides or any other channel forming substances; and
(c) characterizing the effects of lipid environment on the ion permeation and gating of ion channel proteins.

The specific bilayer capacitance, Cb, which reflects thermodynamic forces controlling the thinning of the lipid into a membrane (see Miller, 1986), is used to determine the formation of optimal bilayers.

METHODOLOGICAL APPROACH

Lipid bilayers and vesicle fusion

Bilayers are painted with aid of a fire polished electrode across a 150 μm hole in the wall of a 1 ml polyester cup (Figure 3.1 A) using a mixture of palmitoyl-oleoyl-phosphatidylethanolamine, palmitoyl-oleoyl-phosphatidylserine and palmitoyl-oleoyl-phosphatidylcholine (5:3:2, by volume), in standard lipid extraction solvents, e.g. chloroform and methanol. The lipid in the solvent forms large aggregates organized to shield the polar regions of the lipid from the nonpolar solvent. The lipid mixture is dried under a stream of N_2 and redissolved in *n*-decane at a final concentration of 50 mg/ml. This is because in the presence of the extraction solvent the bilayer undergo time-dependent changes in composition and electrical properties on contact with water. Lipid thinning can be monitored optically through a microscope (a thin lipid bilayer looks black to the eye since it reflects little light) and/or by capacitance measurement. Membrane vesicles are added to the *cis* chamber to a final protein concentration of 1–10 μg/ml and stirred with electronic magnet placed under the *cis* chamber (Figure 3.1: B and C). The fusion is controlled with Ca^{2+} and the osmotic gradients across the vesicles and the bilayer. The cytoplasmic side of the vesicle is thought to face the *cis* chamber (to which membrane vesicles are added).

Recording single channel activity

Typically, the current is measured with an amplifier (see Chapter 5), monitored on an oscilloscope and stored on videotape using pulse code modulation. The *cis* and *trans* chambers are connected to the amplifier headstage by Ag/AgCl electrodes in agar salt-bridges containing the solutions present in each chamber (Figure 3.1 C). Voltages and currents are expressed relative to the *trans* chamber.

Data Analysis

Data are filtered at 1 kHz (4-pole Bessel, –3dB; see Chapter 7) and digitized at 2 kHz. The background noise from bilayers, 150 μm in diameter, has a SD of about 0.8 pA when filtered at 1 kHz. Data are obtained only from bilayers having 0.42 μF/cm^2 which contain one active channel. The criteria for defining and analyzing ion currents that belong to a 'single channel' are described elsewhere (see Chapter 7).

TYPICAL RESULTS

Current components of the bilayer

The voltage ramp (between 0 and –140 mV) induced-current flow, which is used to monitor the state of the black lipid membrane, is shown in Figure 3.2. For an aperture that is covered with lipid film the capacitative current is that of the polyester cup alone I_{Cc} (Figure 3.2 A). After thinning of the lipid film (initially several micrometers thick), the additional capacitative current I_{Cb} is that of the bilayer. Current flow after the change in the bilayer potential has been completed, i.e. dV/dt is zero, is that of the ionic leak, I_{Lb}, of the bilayer. The equivalent electrical circuit of the artificial bilayer (Figure 3.2 B) shows I_{Lb} which is carried by ions crossing the conductive pathway through the bilayer and I_{Cb} which is carried by ions moving up the bilayer to charge or discharge its electrical capacity. The current-voltage relations ($I_{cc} - V_m$) and ($I_{Lb} - V_m$) in Figure 3.2 C are for the cup alone and for the bilayer. Under the experimental conditions reported here a bilayer which has a maximal Cb value of 0.42 μF/cm^2 and a minimum cord conductance for the leak, g_{Lb}, of 12.5 pS at negative voltages is considered optimal.

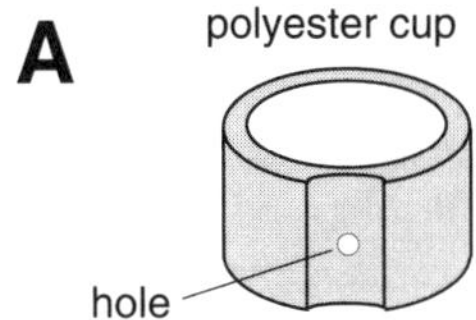

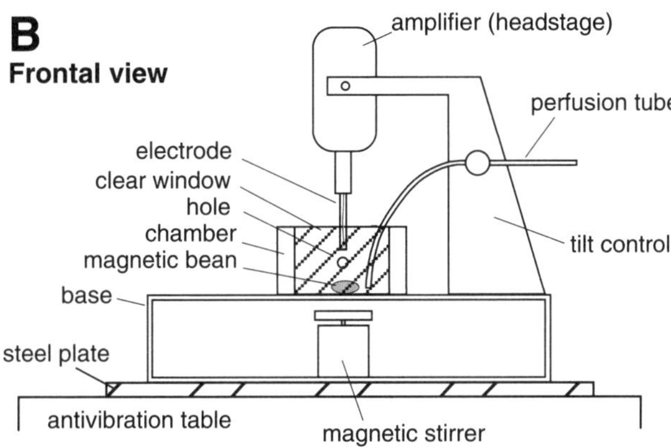

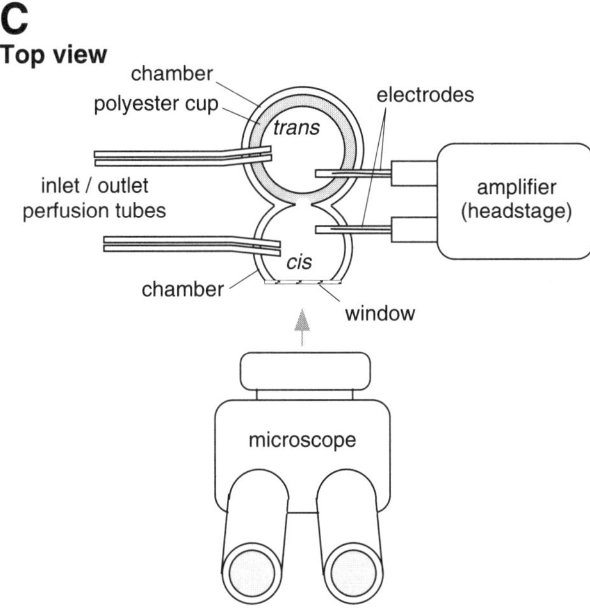

Figure 3.1 A: Polyester cup containing a hole with a diameter of 150 µm. B: Front view of the membrane chambers and current amplifier on a vibration-proof table. C: Top view of the chambers, electrodes, perfusion tubes and microscope.

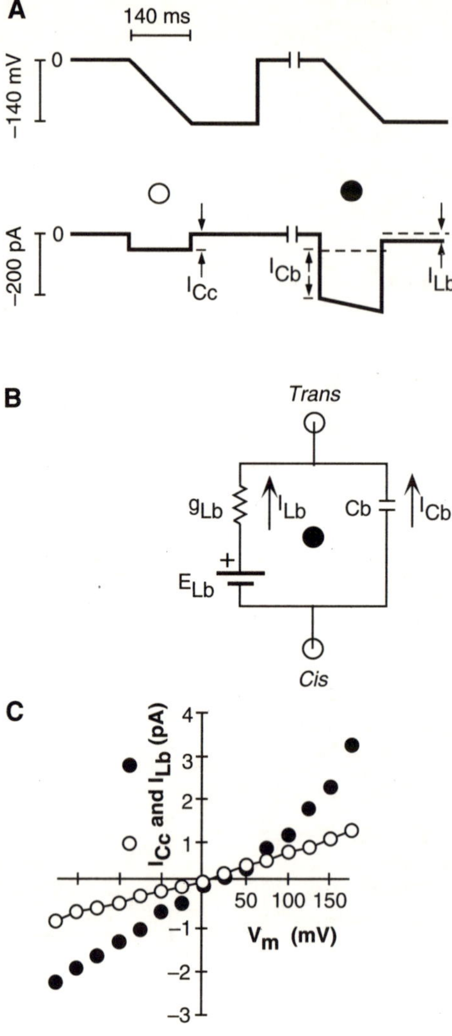

Figure 3.2 A: Voltage ramp (top) from 0 to –140 mV in 140 ms applied across a bilayer spanning the 150 μm aperture in a delrin cup. Current response (bottom). I_{Cc} is the capacitative current of the cup, I_{Cb} is the capacitative current of the thinned bilayer and I_{Lb} is a time-independent leak across the bilayer. B: Equivalent electrical circuit for a bilayer illustrating the passive properties of the bilayer. I_{Lb} with an electromotive force (E_{Lb}) is in series with the leak conductance (g_{Lb}) and in parallel with bilayer capacitance (Cb). C: Current-voltage relations for I_{Cc} and I_{Lb} obtained for a bilayer having a Cb value of 0.42 μF/cm^2. (Reprinted from the *Journal of Membrane Science*, **116**, J. I. Kourie, Vaganes of artificial bilayers and gating modes of the SCI channel from the sarcoplasmic reticulum of skeletal muscle, 221–227, 1996 with kind permission from Elsevier Science—Nl, Sara Burgerhanstraat 25, IO55 KV Amsterdam, The Netherlands.)

Current components of bilayers containing a fused vesicle

In addition to I_{Cb} and I_{Lb} new current components are introduced into bilayers containing a fused vesicle (Figure 3.3). These include (a) I_{iGC} and I_{iNGC} through gated and non-gated channels (Figure 3.3 A) and (b) capacitative current (I_{Cv}) and leak currents (I_{Lv} and I_{Lbv}) of the fused vesicle and its interaction with the bilayer (Figure 3.3 B). For an optimal bilayer I_{Cv}, I_{Lv} and I_{Lbv}, shown in the equivalent electric circuit (Figure 3.3 C), are negligible. In addition, the values of Cb and g_{Lb} are maintained unchanged. The main current in the equivalent electrical circuit for such an optimal bilayer is the variable current I_{iGC}. All bilayers which experience an increase in I_{Lb} after vesicle fusion are rendered not optimal for recording SCl channel activities. The detailed activities of the SCl channel, which is an example of a voltage gated Cl channel, can be seen superimposed on the initial time-independent current I_{Lb} in Kourie (1996).

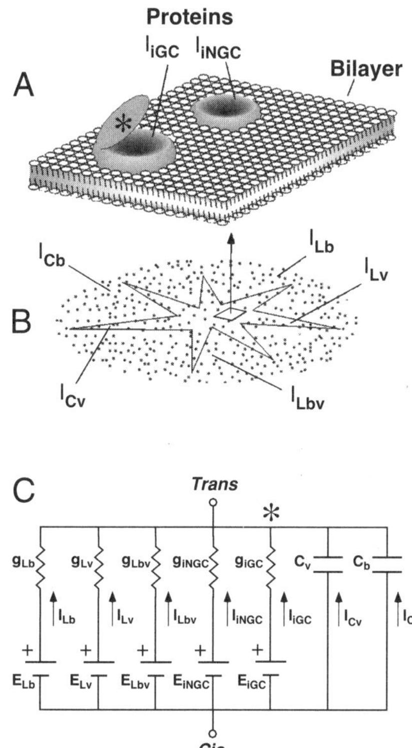

Figure 3.3 Schematic diagram illustrating (A) lipids and channel proteins (B) of a native SR vesicle fused into an artificial bilayer. C: Current flow in the equivalent electrical circuit for a bilayer containing both lipid and channel proteins of a native SR membrane (see text). (Modified from Kourie (1996).)

REFERENCES AND FURTHER READING

Kourie, J. I. (1996) Vagaries of artificial bilayers and gating modes of the SCl channel from the sarcoplasmic reticulum of skeletal muscle. *J. Membrane Sci.* **116**, 221–227.

Miller, C. (1986) *Ion Channel Reconstitution*. New York: Plenum Press.

Mueller, P., Rudin, D., Tien T. H. and Westcott, W. C. (1962) Reconstitution of excitable cell membrane structure in vitro. *Circulation* **26**, 1167–1171.

SECTION 2
ELECTROPHYSIOLOGICAL TECHNIQUES

CHAPTER 4

EXTRACELLULAR SINGLE NEURONAL RECORDINGS IN THE WHOLE ANIMAL

T. R. Vidyasagar
Centre for Visual Science and Division of Neuroscience, John Curtin School of Medical Research, Australian National University

INTRODUCTION

Recording the activity of single neurons in the whole animal is a vital component in the study of integrative functions of the nervous system. One relatively simple way of evaluating such activity without significant disruption of the system is to record the action potentials using a suitable microelectrode inserted in the extracellular compartment. While such recordings do not give information about the inputs to a particular cell (excitatory and inhibitory post-synaptic potentials or currents), as intracellular recordings would, they can provide a wealth of information about the functional properties of single cells and neuronal connectivity in the intact nervous system. Elaborate methods have been developed to study such activity in either anaesthetized preparations or in awake animals without causing pain or distress to the animal. This chapter will give a brief overview of these techniques.

ELECTROPHYSIOLOGY

The potential recorded from an electrode in the extracellular space is influenced by the electrical currents set up by the neuronal elements around it. These elements, be they axons, cell bodies or dendrites, have an electrical potential difference across their cell membranes and these potentials may change with time due to the action of various neurotransmitters that are released at synapses and due to the initiation of action potentials. For example, a pyramidal cell, whose apical dendrites have received excitatory synaptic inputs will produce an active sink for current and the soma will be a passive source, thus constituting an electric dipole. An electrode is surrounded by a number of variable electric dipoles which determine the direction and strength of current flow at the tip of the electrode at every instant. It is usually the aim of extracellular recordings to isolate the activity of just one unit—either the activity of a nerve fiber or that of a cell soma—by the use of an appropriate microelectrode. This is done by bringing the electrode tip close enough to a neuronal element that the signal-to-noise ratio from that one is larger than that from other units that can be recorded in the vicinity.

These electrodes are either glass micropipettes filled with an electrolyte (2 or 3 M, sodium chloride or potassium chloride) with a tip diameter around 1 micron or metal electrodes (usually, tungsten, steel or platinum-iridium) insulated by glass or lacquer with an exposed tip of 5 to 10 μm. An important characteristic of these electrodes is their electrical resistance, which is related to the exposed tip size. Smaller tips have higher resistances and they restrict the area from which potentials can be recorded, thus permitting the isolation of the activity of either a fiber or a cell. Large tips and low resistances pick up the activity from a number of neuronal elements and are of limited use in efforts to identify the functional properties of single nerve cells. Tips with very high resistances are also of little use since they cannot record the neural activity unless they are very close to the neuronal cell membrane or actually inside a cell. Electrodes often also have significant capacitances at the metal/fluid interface and due to the thinness of the isolating glass or lacquer near the tip. Usually, the impedance (which is a measure of electrical resistance plus capacitance) of the most successful electrodes is in the range of 5 to 20 MOhms when measured by an AC current at 50 Hz. Electrodes with impedances less than 3 or 4 MOhms tend to record from more than one cell simultaneously. The spikes are smaller and usually only a multi-unit recording can be made. While the earliest electrodes used were glass micropipettes, metal electrodes are more commonly used nowadays, especially a variety of tungsten-in-glass electrodes (Hubel, 1957; Levick, 1972; Merrill & Ainsworth, 1972; Vidyasagar & Perry, 1979). These metal electrodes not only provide more stable isolation of single units than micropipettes, but tend to sample from a larger morphological variety of cells and also help in better localization of electrode tracks (see below).

Over the last fifty years, extensive extracellular recordings have been made from almost every part of the brain in a number of animal species and even in human beings in a limited way during neurosurgical operations. When the electrode is inserted into the brain tissue, unit activity is encountered under one of four conditions: (a) some neurons are spontaneously active; (b) the electrode tip may injure a neuron and provoke it into firing; (c) the neuron may fire in response to a natural stimulation, such as a visual or an acoustic stimulus; or (d) the neuron may fire due to an artificial stimulus, such as an electrical stimulus applied elsewhere, or due to a chemical stimulus applied in the close vicinity of the cell.

In most extracellular studies, it is the frequency of neuronal spikes that is of most interest and not necessarily the shape of the action potential. Therefore, the recordings are usually done with an AC amplifier, with which most irrelevant signals such as DC shifts and high frequency noise can be filtered out. These amplifiers usually have filters to cut off frequencies below and above a particular range (band-pass filters), and allow an amplification of at least 10 000 times the actual signal. The most convenient range for the filter that optimizes spike recording without allowing too much noise is 300–6000 Hz. However, if the study requires attention to spike shape, it is better to widen the filter at the low frequency end or use DC recording as for intracellular recordings. The signal is normally visualized on an oscilloscope. With good earthing systems and an appropriate amplifier, the noise level observed in AC recordings should be below 100 μV. The output of the amplifier as well as the output of the window discriminator (see below) are also fed into a loudspeaker, so that the experimenter can 'listen' to the electrical activity picked up by the microelectrode.

Generally, fiber spikes, such as those from the optic tract, and soma spikes, like those from the principal cells of the thalamus, differ markedly in their shapes and how these change as the electrode approaches them (Hubel, 1960; Bishop et al., 1962). Fiber spikes generally appear as positive potentials, primarily because current flowing either in front of, or behind, an action potential is outward current (source). On the other hand, soma spikes can first appear as either negative or positive waves depending upon where the electrode is relative to current sinks and sources. When an electrode approaches the soma of a neuron which is undergoing synaptic activation the initial wave will be small and positive as current flows into the dendrites. This will be followed by a negative wave due to inward current in the soma as it is invaded by the depolarization of an action potential (Figure 4.1 A). A rapid increase in the amplitude of the positive potential occurs as the electrode gets close to the cell surface. Sometimes there are inflections on the ascending potential, indicating that the somatic membrane is acting as a current source for depolarisation in dendrites and in the initial segment (Figure 4.1 B). When the electrode touches the cell body membrane and causes damage locally, there might be deterioration of the recording with the negative wave often dropping out entirely (Figure 4.1 C). There

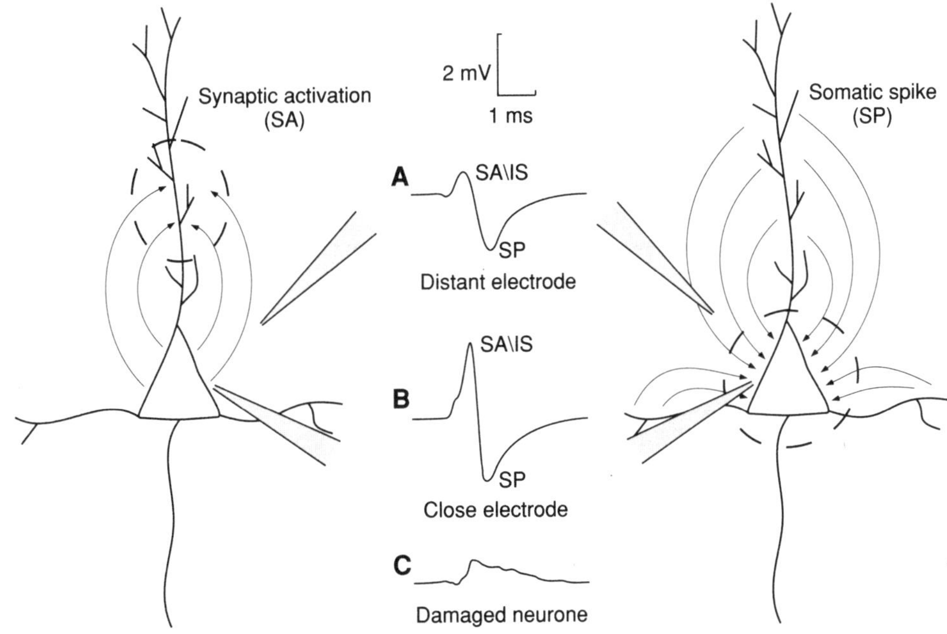

Figure 4.1 A simplified schematic diagram of the sources and sinks that are responsible for the potentials recorded by an extracellular electrode as it approaches the soma of a pyramidal cell. The neuron is shown in two states: first, at the stage where depolarisation is occurring in the apical dendrites (on the left), and, second, when the soma itself is undergoing depolarisation (on the right). The waveform in B is produced by the electrode being much closer to the cell body than in A. The negative waves seen in both A and B represent the sink around the cell body and the early positive waves from the source for depolarisation occurring in dendrites and/or the axon initial segment (IS). Waveforms as in C are often seen when the membrane has been damaged by the electrode tip pushing on it.

have been a number of early accounts that deal at length with the theoretical basis of potentials in a 'volume conductor' that is seen by a recording electrode (Ruch & Patton, 1965; Burěs, 1967; Plonsey, 1969). However, the precise geometry of the dipoles around the electrode, their variation with time and the conductivity of the extracellular medium all affect the electrical currents in the tissue and little more than a very general, qualitative description can be made of the events. Furthermore, the filters that are sometimes used during extracellular recordings lead to differentiation of the signal and such filtered signals would appear different from those displayed in Figure 4.1.

One common difference between fiber and soma recordings is the overall duration of the spikes, which is usually much less than 1 ms for fibers, but could be up to 2 ms or more for cell bodies. However, in regions where different cell types and axons may be encountered, caution is necessary in relying on spike shape for classifying fiber and soma spikes. This is because, some cells, in particular the small inhibitory interneurons (non-spiny stellate cells) in the cortex, have spikes that are not very different from those of fibers (McCormick et al., 1985). However, it appears that there are reliable differences between the spikes of these interneurons, called fast-spiking cells, and those of the pyramidal neurons, described as regular spiking or bursting (for details, see McCormick et al., 1985).

ACUTE ANIMAL PREPARATIONS

Detailed protocols have been established for stable single unit recordings from the brains of anaesthetized animals for periods lasting up to 3 to 5 days (eg., Hubel & Wiesel, 1962; Bishop et al., 1971). The animals are anaesthetized during the entire procedure and it is important that stable levels of anaesthesia and an excellent physiological state are maintained to enable reliable unit recordings. Usually, the animals are also paralysed so that controlled artificial ventilation can be undertaken. For lengthy experimentation it is essential that scrupulous attention is paid to optimal levels of ventilation, which can only be achieved with full skeletal muscle paralysis. Such paralysis may also be necessary in experiments on the visual system to eliminate eye-drifts, that may otherwise interfere with receptive field plotting. Concentration of carbon dioxide in the expired air is continuously monitored to help in adjusting the rate and tidal volume of ventilation. The expired CO_2 is usually kept between 3.6 and 3.8% for mammals such as cats and monkeys, to provide optimum alveolar ventilation. The electrocardiogram (ECG), subscapular or rectal temperature and electroencephalogram (EEG) are also continuously monitored. The heart rate and the EEG also provide important feedbacks regarding the adequacy of anaesthesia. Most often, the anaesthesia used during the experiment is a combination of a gaseous anaesthetic such as nitrous oxide or halothane and an intravenously administered barbiturate. A continuous intravenous infusion of glucose and electrolytes is also maintained.

A craniotomy is performed using a dental drill and the electrode is introduced through a small hole made in the dura mater. Stability, i.e. lack of any movement between electrode and brain tissue, is vital for achieving single unit recording. For this, the animal is usually fixed in a stereotaxic frame on which a stable electrode carrier ('microdrive') is mounted to carry the electrode (see Chapter 15). Brain pulsations are often a problem in recordings from the cortex. To dampen these, various procedures have been used, such as a sealed chamber over the craniotomy or simply pouring an agar solution into a well, built around the craniotomy. Further stability can be achieved by performing bilateral pneumothorax, since variations in the intrathoracic pressure contribute significantly to the brain movements.

For identification of the electrode tracks, electrolytic lesions are made with DC current (say, 6 μA for 6 s) through the metal electrode. At the end of the experiment the animal is given a lethal dose of the anaesthetic and perfused with a formaldehyde solution through the heart. After appropriate processing of the brain sections of required thickness (say, 50 μm) are made and stained for Nissl substance. In these sections, the lesions can be identified, the electrode tracks reconstructed and the position of the recorded units located with respect to cortical area and lamina. There are, however, many different histological methods that could be combined with the physiology, depending upon the scientific question that is being investigated.

CHRONIC ANIMAL PREPARATIONS

Working on anaesthetized preparations, as outlined above, has the limitation that higher brain functions that require a fully conscious animal cannot be adequately studied. Therefore, special procedures have been developed to record single cell activity from the brains of awake animals without causing them pain or discomfort. Much of this type of work is carried out on monkeys, since they can be easily trained on complex behavioral tasks and when given appropriate reward, readily co-operate in performing these tasks.

The method involves the following steps (for more details see, Lemon, 1984; Vidyasagar, 1993):

1. The monkey is first trained on one or more specific paradigms. For example, it is common in experiments on the visual system that the monkey is trained to fixate, to discriminate between different visual patterns or to perform some specific pattern of eye movements. The actual detail depends upon the question being investigated. In experiments on memory, it may be trained on a 'delayed match-to-sample' task, where the monkey is required to remember a sample pattern for a certain period and then match it to one of several different test patterns.
2. After a certain level of performance has been reached a surgical operation is performed under general anaesthesia and aseptic conditions. A small opening is made into the skull over the brain area of interest and a chamber with a screwable cap is then implanted over it.
3. When the animal has recovered from the operation, electrophysiological recordings are done for a few hours each day with an electrode mounted on a lightweight microdrive mounted on the chamber. Since brain tissue itself is devoid of pain receptors

the electrode penetration is painless. However, attention needs to be paid to keep the chamber itself clear of infections. This requires regular cleaning of the chamber and the use of appropriate local antibiotics at the earliest sign of any infection.

The recording of unit activity during the performance of the trained tasks helps in directly identifying the neural basis of various cognitive and behavioral phenomena. This is one of the few techniques available for investigating complex brain processes at a functional level.

DATA ANALYSIS

In most extracellular studies the signal parameter of interest is the spike frequency and how it varies with time and with respect to the applied stimuli. If the electrode picks up the activity of more than one unit, they can still be discriminated by exploiting the fact that these spikes from different neuronal elements usually have different heights. To record the time of occurrence of the spikes of a certain height and thus the activity of one neuron, the analog signal from the amplifier is fed into a suitable trigger device, usually a 'window discriminator'. This device produces a single pulse of given height and duration (known as a TTL pulse) for each action potential whose height is within a certain window. The output of the window discriminator is usually fed into a computer and a dedicated software compiles the data in a suitable form. Computer programs are now available that can identify spike templates for each of the units in a multi-unit recording and process separately the response of each unit (eg. Brainwave Systems' Discovery Package).

One common method of displaying the data obtained from extracellular recordings is a peristimulus time histogram (PSTH), where a stimulus (e.g. a moving or a flashing light bar) is presented many times and the spikes produced by a neuron in the visual system are recorded at the same time by the computer (Figure 4.2). The X-axis in the figure represents time, divided into a number of bins of a certain duration, and the Y-axis represents the number of spikes that occur in each of the bins, most often expressed in spikes/sec. Repeated presentation of a stimulus and averaging of the response is necessary to eliminate the effects due to variability in response and neuronal excitability that are unrelated to the stimulation. When investigating the neuronal response to a stimulus variable (such as the velocity of a moving visual stimulus), it is necessary to interleave the presentation of the different stimuli to eliminate spurious results due to response variability (Henry et al., 1973).

CONCLUSION

Apart from studying the function of the single cell in question, extracellular recordings have been used in many ingenious ways to investigate a variety of related issues. For example, electrical stimulation can be applied in one area and the spike response recorded

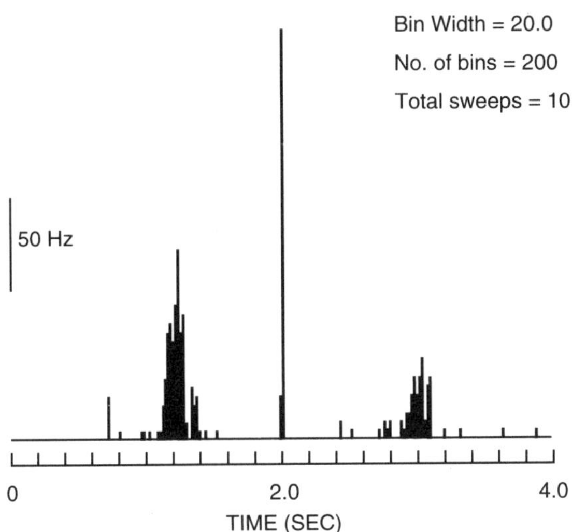

Figure 4.2 A typical peristimulus time histogram (PSTH). The record taken is the response of a simple cell in the primary visual cortex (area 17) of the cat to ten repeated sweeps of a bar (8° long and 0.2° wide) across the receptive field in both directions. The best stimulus orientation was 40° anticlockwise to the vertical. During the first half of the PSTH (i.e. the first 2 s), this bar was moving from bottom left to top right of the screen and during the second half in the opposite direction. The tall line in the middle (at 2 s) denotes the point where the bar reverses the direction. The velocity was 2°/s and the bar was four times brighter than the background. The sweep in each direction is divided into 100 bins of 20 ms each and the spikes that occur in each bin are added together to create the histogram. Note that the cell responds much better in the first direction, i.e. the cell exhibits 'directional selectivity'. The response was recorded through a tungsten-in-glass microelectrode. The electrode had a multi-barrelled electrode glued to it, the barrels of which contained different drugs. The particular response shown here was recorded while bicuculline methiodide, an antagonist of the inhibitory transmitter, GABA, was being ejected through one of the barrels. Before the application of the drug (not shown here), the cell was not responding to the first direction at all. Removal of the inhibition by bicuculline reveals that the original lack of response for the first direction was due to GABA-mediated inhibition.

in a different area. These studies not only demonstrate the connections between brain regions, but also help to reveal the nature of such connectivity. The activity of two different neurons, either from the same electrode or from two separate electrodes, can be cross-correlated to see whether one neuron is influenced by the firing of the other. Such cross-correlograms done between two or more neurons are giving new insights into the integrative functions of the nervous system. In experiments lasting many days, extracellular recordings can also be used to rapidly identify basic functional properties of a very large number of neurons in one or more brain regions to study the principles of overall topographical organisation. Thus, extracellular single neuronal recording has remained a powerful tool in investigating neuronal function in the intact animal.

RECENT PAPER

Colby et al. (1996) extracellulary recorded the neuronal responses in a part of the awake monkey's parietal cortex as the animal performed a number of cognitive tasks. These involved controlled eye movements in response to visual stimuli of specific types and directed attention to cued locations. The study is a typical example of experiments done on conscious animals and uses a number of methods described in this chapter. There are also many figures depicting PSTHs (Figures 2, 5, 7 and 9) and related ways of processing data.

REFERENCES AND FURTHER READING

Bishop P. O., Burke W. and Davis R. (1962) The identification of single units in central visual pathways. *J. Physiol.* **162**, 409–431.

Bishop, P. O., Henry, G. H. and Smith, C. J. (1971) Binocular interaction fields of single units in the cat striate cortex. *J. Physiol.* **216**, 39–68.

Burěs, J., Petrán, M. and Zachar, J. (1967) *Electrophysiological Methods in Biological Research*. London: Academic Press.

Colby, C. C., Duamel, J-R. and Goldberg, M. E. (1996) Visual, presaccadic, and cognitive activation of single neurons in monkey lateral intraparietal area. *J. Neurophysiol.* **76**, 2841–2852.

Henry, G. H., Bishop, P. O., Tupper, R. M. and Dreher, B. (1973) Orientation specificity and response variability of cells in the visual cortex. *Vision Res.* **13**, 1791–1799.

Hubel, D. H. (1957) Tungsten microelectrode for recording from single units. *Science* **125**, 549–550.

Hubel, D. H. (1960) Single unit activity in lateral geniculate body and optic tract of unrestrained cats. *J. Physiol.* **150**, 91–104.

Hubel, D. H. and Wiesel, T. N. (1962) Receptive fields, binocular interaction and functional architecture in the cat's visual cortex. *J. Physiol.* **160**, 106–154.

Lemon, R. (1984) *Methods for Neuronal Recording in Conscious Animals*. New York: Wiley.

Levick, W. R. (1972) Another tungsten microelectrode. *Med. Biol. Eng.* **10**, 510–515.

McCormick, D. A., Connors, B. W., Lightall, J. W. and Prince, D. A. (1985) Comparative electrophysiology of pyramidal and sparsely spiny stellate neurons of the neocortex. *J. Neurophysiol.* **54**, 782–806.

Merrill, E. G. and Ainsworth, A. (1972) Glass-coated platinum plated tungsten microelectrodes. *Med. Biol. Eng.* **10**, 662–672.

Plonsey, R. (1969) *Bioelectric Phenomena*. New York: McGraw-Hill.

Ruch, T. C. and Patton, H. D. (1965) *Physiology and Biophysics*, (19th edn). Philadelphia: Saunders.

Vidyasagar, T. R. (1993) Assessment of brain electrical activity in relation to memory and complex behavior. *Meth. Neurosci.* **14**, 407–431.

Vidyasagar, T. R. and Perry, G. W. (1979) An improved simple tungsten microelectrode. *Brain Res. Bull.* **4**, 285–286.

CHAPTER 5

EQUIPMENT FOR MEASUREMENT OF MEMBRANE POTENTIALS AND CURRENTS, WITH SPECIAL REFERENCE TO SHARP MICROELECTRODES

R. L. Martin
Division of Botany and Zoology, The Faculties, Australian National University

INTRODUCTION

Conventional equipment for recording of membrane potentials in research laboratories includes a microelectrode, headstage, micromanipulator, amplifier (electrometer), oscilloscope, a timing device and usually a loudspeaker system (Figure 5.1). These days some research laboratories are using computers to replace the timing device and oscilloscope (e.g. MacLab system); with appropriate software the computer screen can emulate an oscilloscope or chart recorder. Thus, it is unnecessary to have all the equipment shown in Figure 5.1.

Briefly, the biological signal is passed from the preparation to the headstage via the microelectrode and displayed on an oscilloscope. Most often a permanent record of the data is obtained by converting it from an analog to a digital signal and capturing it on computer. It may also be permanently recorded on magnetic or digital tape. Chart recorders may be useful for long-term monitoring of membrane potential; their further use is limited by their inability to follow very fast events such as action potentials. In the following sections each of the main pieces of apparatus will be discussed in turn.

RECORDING AND REFERENCE MICROELECTRODES

For intracellular recording a sharp microelectrode penetrates the cell membrane and the membrane potential is the difference between the intracellularly recorded potential and that recorded by a reference electrode located outside the cell (Figure 5.1). Ideally the reference electrode would be placed immediately outside the penetrated cell. However, this is usually impossible and therefore the reference electrode is placed nearby, e.g. within the surrounding tissue of a brain slice or under the scalp if intracellular recordings are being made from the brain *in vivo*. Sometimes the reference electrode is at, or near, zero potential, as shown in Figure 5.1, by connection to earth. It is often then referred to as a 'ground electrode'. The reference electrode may be a silver wire or it may be a glass microelectrode filled with the same electrolyte used in the recording electrode but made up in agar (to stop the electrolyte leaking from the tip as it may affect the biological tissue). The choice of reference electrode is important because a potential difference, called a 'junction' or 'diffusion' potential, exists between any liquid–liquid or metal–liquid junction. It is dependent upon the mobilites of the cations and anions in

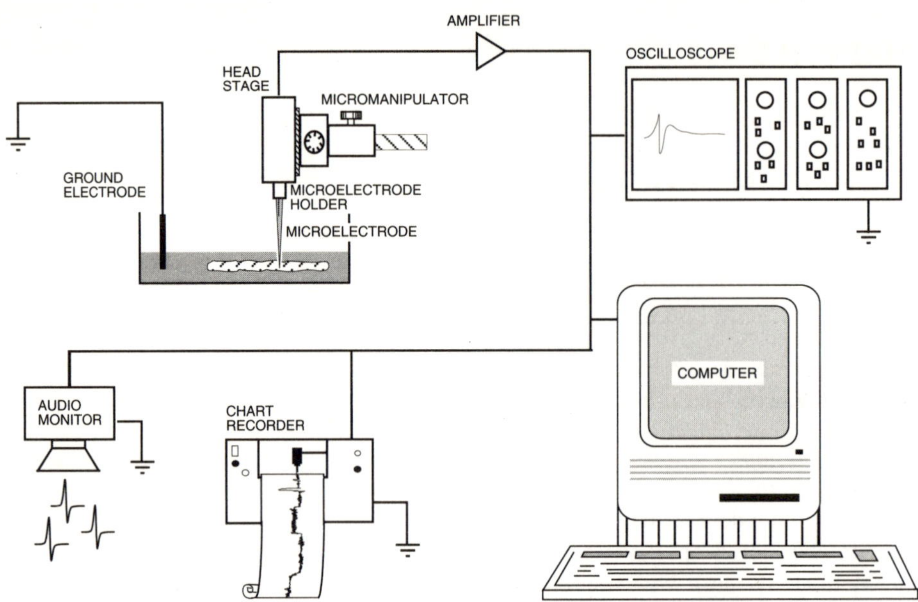

Figure 5.1 Schematic diagram of equipment often used for intracellular recording with sharp microelectrodes. Similar pieces of equipment are used for whole-cell recording.

solution and concentrations of the ions at the interface. Junction potentials occur whenever the cations and anions have different mobilities and the electrolytes at the junction have different concentrations. One reason that potassium chloride (KCl) is often used as an electrode filling solution is that K^+ and Cl^- have similar mobilities in aqueous solution which results in a small junction potential. While it is not possible to eliminate junction potentials, if those for the recording and reference electrode are similar and stable, then they present little problem for sharp microelectrode recording. However, with large patch electrodes and whole-cell recording (see Chapter 6) junction potentials can be significant and must be taken into account (Neher, 1995).

Sharp microelectrodes for intracellular recording are typically fine-tipped (<1 μm) tapering glass pipettes, like that schematically depicted in Figure 5.1. Electrical continuity between the neuron which has been penetrated and the headstage is provided by filling the microelectrodes with a strong electrolyte such as potassium chloride (often 3 M), potassium acetate or potassium methylsulphate. The choice of filling solution depends on the nature of the experiment being undertaken, e.g. studies requiring a normal intracellular Cl^- concentration would not usually be undertaken using a potassium chloride filled microelectrode because small amounts of solution continually leak out of the microelectrode tip into the neuronal interior, and thus the Cl^- concentration inside the cell may not be stable. Microelectrodes are manufactured from capillary tubes using a microelectrode puller, a device which heats the glass and pulls it apart in a controlled manner (see e.g. Brown & Flaming, 1986).

These days relatively few investigators use sharp microelectrodes. Instead, blunt, large-tipped electrodes are used to patch onto the neuronal membrane. More details of this approach are given in Chapters 6.

MICROELECTRODE HOLDER, HEADSTAGE AND MAIN AMPLIFIER

The microelectrode is placed in the microelectrode holder; this usually has a Ag/AgCl pellet in its upper end to provide electrical continuity between the microelectrode filling solution and the wires which convey the signal. The holder has a plug which allows it to fit neatly into the headstage and in doing so it provides electrical continuity. The headstage is held by a micromanipulator, a device which allows fine movements in the X-, Y- and Z-axis, and thus permits precise positioning of the microelectrode in tissue.

Although microelectrodes are typically filled with highly ionic solutions they still offer considerable resistance to current flow (from 5–150 $\times$ 10^6 Ohms; this can be checked by measuring the voltage drop across the microelectrode when a known current is passed down it). If a typical microelectrode is connected to a headstage which itself has only low input resistance then the measured membrane potential is grossly underestimated (see Johnston & Wu, 1994). To avoid this problem and ensure that the output of the headstage follows the voltage at the microelectrode tip, the headstage has a very high input resistance (about 10^{12} Ohms).

The signal is passed from the headstage to the main amplifier where amplification of the signal takes place (up to 100×). In addition, the main amplifier often allows the experimenter to choose the mode of recording, e.g. current-clamp, discontinuous current-clamp or voltage-clamp, to compensate for effects arising from microelectrode and 'stray' capacitance, and to compensate for any non-biological DC offsets which occur in the recording set-up (DC offset).

Current-clamp mode is usually chosen for direct measurement of membrane potential. In this mode the experimenter can inject steady-state current (whose amplitude can be set by the experimenter) into a neuron to change membrane potential. Because the current through the microelectrode is steady this method of obtaining measurement of membrane potentials is called 'current-clamp'. Special circuitry is required to ensure that all the current generated by the current source in the main amplifier box does indeed flow into the neuron. Once again, correct measurement of membrane potential requires that any voltage drop associated with passage of current through the microelectrode is compensated. This is achieved by subtracting a signal which is proportional to the product of the microelectrode resistance and current from the output of the headstage amplifier. Historically, the circuitry required to do this was known as a Wheatstone Bridge and thus, the name 'bridge balance', rather than current-clamp, is used by some researchers. The correct use of the modern equivalent of the Wheatstone Bridge permits the experimenter to directly obtain a measure of the microelectrode resistance from a dial on the main amplifier.

In discontinuous current-clamp mode the single microelectrode switches cyclically between passing current and recording voltage. The experimenter sets the cycle frequency which is usually in the range 5–30 kHz (cycle durations of less than 200 μs).

The microelectrode passes current for about 30% each cycle and in the remainder of the cycle the voltage drop caused by passage of the current through the tip of the microelectrode dissipates according to the time constant of the microelectrode (i.e. product of the microelectrode's resistance and capacitance). Late in the cycle a true measure of membrane potential can then be made without being confounded by the microelectrode resistance. In effect, measurement of membrane voltage takes place in a very small part of each cycle.

Microelectrodes exhibit considerable transmural capacitance when immersed in one or two millimetres of fluid. In addition, 'stray' capacitance arises between the microelectrode and grounded surfaces, from the microelectrode holder, and from the capacitance between the headstage amplifier and grounded surfaces. These capacitances can very effectively change the nature of the recorded signal such that low frequencies are faithfully recorded whereas high frequencies are filtered out. A compensatory circuit is included in the main amplifier box which is called 'capacitance neutralization'. It is always best to try to minimize the capacitance, for example, by minimizing the depth of fluid in which the microelectrode is immersed or by pulling microelectrodes from thick-walled glass (since capacitance, C, is proportional to the surface area divided by the thickness of the glass).

The DC offset dial of the main amplifier box has an important role in compensating for junction potentials. These potentials are substracted by the offset control which allows the measured potential to be set to 0 mV immediately before attempting an intracellular penetration.

OSCILLOSCOPE AND COMPUTER ACQUISITION

The cathode ray oscilloscope is the device used for measuring the amplitude of the voltage signal in electrophysiology. It receives the signal from the amplifier and displays as a function of time. In other words, the oscilloscope really displays a graph of y against x where y is voltage and x is time. One important point about the oscilloscope is that it can measure changes in voltage very accurately and can capture very fast events, e.g. events that occur on the millisecond time scale. Some degree of amplification is also available, as may be filtering.

As mentioned earlier, some laboratories are doing without oscilloscopes and instead using computers to monitor and record biological signals. There are now a number of commercially available software packages which allow a computer to emulate an oscilloscope. Additionally, experimental protocols can be controlled by use of appropriate software, and computers can offer sophisticated data analysis and storage of vast quantities of data.

TIMING DEVICE AND AUDIOAMPLIFIER

In many experiments it is necessary to observe the neuronal response repeatedly, e.g. axons may be repetitively stimulated with an applied voltage and the post-synaptic response averaged, or short pulses of current may be passed down a microelectrode to observe the action potential and ensure the microelectrode penetration is neuronal rather than glial. Essentially the experimental protocol is repeated over many cycles. A timing device allows this to be easily accomplished because not only can it be used to set the cycle duration but it can also be used to set the duration and time for delivery of a stimulus or current pulse. The output of the timing device is then connected to a stimulator or to the main amplifier, or perhaps to both. The timing device can also be used to trigger the oscilloscope, thereby initiating an oscilloscope sweep just before delivery of a stimulus or a current pulse. This ensures that the biological response is easy to see because it will occur at the same place on the oscilloscope screen in every cycle. Increasingly, computers and appropriate software are replacing timing devices.

An audioamplifier is often part of an electrophysiological set-up because the production of a sound, whose frequency is related to the voltage being recorded, makes it possible to 'hear' what is happening at the tip of the microelectrode. A change in sound alerts the experimenter to the fact that the microelectrode is in close proximity to a firing neuron.

VOLTAGE-CLAMPING

Discontinuous single electrode voltage-clamping (DSEVC)

Many ionic channels of the neuronal membrane are voltage-dependent, i.e. the conductance of the channels varies as membrane potential changes. To study the properties of such channels it is usually necessary to make controlled step changes in membrane potential and *measure the transmembrane current* that results. Of course there are numerous other experimental situations when it may be ideal to control membrane potential, e.g. to determine the reversal potential of a synaptic potential.

The general principle can be described quite simply. Membrane potential is measured at the tip of the sharp microelectrode and compared with the value that the experimenter 'commands'. The difference between measured membrane potential and the command potential is obtained and current sufficient to force membrane potential to equal the command potential is passed via a microelectrode into the neuron. If channels in the membrane open as a result of the imposed voltage step then the voltage-clamp must pass current which is equal and opposite to that through the channels in order to hold membrane potential at the command value. Thus, a measurement of transmembrane current is obtained which reflects the underlying membrane conductances, i.e. the number of open channels. This is indeed the case because capacitive current only flows transiently when membrane potential is changed ($I_C = C\ dV/dt$). This capacitive current can be seen at the start of the current trace.

Historically, voltage-clamp experiments were undertaken using two intracellular electrodes, one measuring membrane potential and the second measuring membrane current. But this is only possible where the cell of interest is sufficiently large to accommodate two electrodes without causing significant damage. The neurons of the CNS are generally not large and therefore voltage-clamp methods were modified in such a way that a single electrode could carry out the dual functions of measuring membrane potential and membrane current. Essentially, switching between these two functions occurs in the same way as that described above for discontinuous current-clamp. The difference between the command voltage and membrane potential is assessed when the voltage drop across the microelectrode's tip caused by passage of a brief pulse of current has dissipated. Current sufficient to hold membrane potential at the command voltage is then passed (this may be expressed as current per mV of voltage error). Strictly speaking this approach is called discontinuous voltage-clamp, a name which reflects the fact that the electrode measures membrane potential and passes current at different moments in time.

Vigilant experimenters can avoid some problems associated with voltage-clamping using sharp electrodes. The high resistances of these electrodes means they may not readily pass all the current from the amplifier. Consequently, membrane potential may not be clamped at the command potential. However, this is readily detected by continuously monitoring membrane potential (as well as the command potential). Another problem is that the voltage-clamp may not be fast enough to fully clamp very fast changes in membrane potential. An assessment of the speed of the clamp can be made from the 10–90% rise time of voltage step applied at the soma.

Continuous single electrode voltage-clamping (CSEVC)

With this technique one electrode is used to simultaneously measure voltage and pass current; it is commonly referred to as whole-cell patch-clamp. A relatively large-tipped patch electrode is used to seal onto the membrane and then the membrane patch under the electrode is ruptured (see Chapter 6). This provides a relatively low resistance pathway between the patch electrode and the cell interior that is referred to as the 'access' or 'series' resistance. The electronic circuitry is such that the voltage at the headstage end of the electrode is controlled (a very different situation from voltage-clamp with a sharp electrode where the voltage at the tip of the microelectrode is controlled). As a result, the measurement of voltage reflects the sum of membrane potential and the voltage drop across the access resistance during passage of current. Therefore, in CSEVC the actual membrane potential will differ from the command potential by the product of the access resistance and clamp current. Access resistance also influences the time constant, τ, of the measured membrane current because τ approximates the product of access resistance and pipette capacitance (assuming the membrane capacitance is much larger) (Sherman-Gold, 1993). Given the problems associated with access resistance, it is not surprising that most whole-cell patch-clamp amplifiers contain a means of compensating for it. However, rarely can more than 80% compensation be achieved. Clearly, adequate voltage-clamping can only be achieved by maintaining a low access resistance at all times; since ruptured membranes have a tendency to seal over this means that access resistance should be continuously monitored. The residual error from uncompensated access resistance must always be taken into account when interpreting data derived from whole-call patch-clamp experiments.

Pitfalls of all types of voltage-clamping

A number of the pitfalls associated with DSEVC and CSEVC have already been mentioned. The 'space-clamp' problem is a problem associated with both and it cannot be avoided by technical wizardry. Long neuronal processes have axial resistances, the membrane is leaky and it behaves as a capacitor. These properties make the voltage at the end of the dendrites very different (smaller and slower) from the voltage at the tip of the microelectrode. As a result, the dendritic membrane potential is not well controlled by the voltage-clamp. Those parts of the neuron which only experience a portion of the voltage command will generate a current which is an underestimate of the true biological current for the applied command voltage.

RECENT PAPER

Chandler et al. (1994) used sharp microelectrodes to study the properties of guinea pig trigeminal motoneurons *in vitro*. To demonstrate the morphology of these neurons they filled their electrodes with a mixture of KCl and biocytin, an anatomical marker (see Chapter 18). Electrophysiological measurements were made in bridge balance or discontinuous current- or voltage-clamp modes. The methods section of the paper describes the sampling rate and length of the duty cycle when the discontinuous mode was selected and shows careful attention to the moment of sampling through monitoring of the headstage voltage. Figure 2 illustrates both current-clamp and voltage-clamp recordings; note that either the size of the current steps applied to the neuron (current-clamp) or the changes in membrane potential (voltage-clamp) are shown. In voltage-clamp, small deviations of the membrane potential from step changes indicate that it was not always perfectly controlled (e.g. Figure 8A).

REFERENCES AND FURTHER READING

Brown, K. T. and Flaming, D. G. (1986) *Advanced Micropipette Techniques for Cell Physiology*. Chichester, UK: John Wiley & Sons.

Chandler, S. H., Hsaio, C-F., Inoue, T. & Goldberg, L. J. (1994) Electrophysiological properties of guinea pig trigeminal motoneurons recorded *in vitro*. *J. Neurophysiol.* **71**, 129–145.

Johnston, D. and Wu, S. M. (1994) *Foundations of Cellular Neurophysiology*. Massachusetts, USA: MIT Press.

Jones, S. W. (1990) Whole-cell and microelectrode voltage clamp. In *Neurophysiological Techniques: Basic Methods and Concepts* (Neuromethods, Vol. 14), edited by A. A. Boulton, G. B. Baker, and C. H. Vandersolf. New Jersey, USA: Humana Press.

Neher, E. (1995) Voltage offsets in patch-clamp experiments. In *Single-Channel Recording* (2nd edn), edited by B. Sakmann & E. Neher. New York: Plenum Press.

Sherman-Gold, R. (ed) (1993) *The Axon Guide for Electrophysiology and Biophysics Laboratory Techniques*. Foster City, USA: Axon Instruments.

CHAPTER 6

SUCKING UP TO CELLS: THE PATCH-CLAMP TECHNIQUE IN NEUROSCIENCE

J. M. Bekkers
Division of Neuroscience, John Curtin School of Medical Research, Australian National University

INTRODUCTION

Advance in neuroscience is often techniques-driven. One of the best illustrations of this principle is provided by the story of the patch-clamp technique. Introduced in a primitive form in 1976 and refined in 1981, patch-clamping has lead to an explosion of experiments that were previously unimaginable (Sakmann & Neher, 1995). In recognition of this, its inventors, Bert Sakmann and Erwin Neher, were awarded a Nobel Prize in 1991.

Patch-clamping is, in some sense, the modern substitute for microelectrodes, although some things are still done better with microelectrodes (see Chapter 5). Patch-clamping goes much further, however, in permitting whole new classes of experiments that would never have been possible with traditional recording techniques. Some of these kinds of experiments will be described later in this chapter. First, however, some of the basic theory and practice that is common to all patch-clamping will be presented.

PATCH-CLAMP THEORY AND PRACTICE

The basic idea behind patch-clamping is illustrated in Figure 6.1. Unlike traditional microelectrode techniques, where the point is to impale the cell, patch-clamping involves sealing a glass microelectrode against the *surface* of the cell. Accordingly, the tip diameter of a patch electrode is much larger than that of a typical intracellular electrode ($\approx 1\ \mu m$, as opposed to $\approx 0.1\ \mu m$). Furthermore, the tip of a patch electrode is often melted slightly ('polished') to smooth out any sharp edges that might puncture the cell membrane.

The sequence involved in getting a patch-clamp recording is shown in Figure 6.1. First, the patch electrode is simply pressed against the cell membrane using a micromanipulator (A). The cell is bathed in Ringer solution, the electrode is filled with the same or a different Ringer, depending on the experiment (see later), and a sensitive amplifier (the patch-clamp amplifier) is used to record the current flowing through the electrode (C). Gentle suction is then applied (usually by mouth) via a tube attached to the back of the patch electrode. This causes the cell membrane to balloon up into the electrode and, by a process that is not understood, to seal tightly to the glass rim at the tip of the electrode (B). The benefit of this tight seal is understood from elementary noise theory.

The situation in Figure 6.1 A can be reduced to an equivalent electrical circuit. Current flowing down the patch electrode can pass either through the small patch of cell membrane spanning the electrode tip (with electrical resistance R_{patch}), or else it can leak out under the rim of the electrode tip where it contacts the cell membrane (with resistance R_{leak}). Usually $R_{leak} << R_{patch}$, so most of the current goes via the leak. In this case it can be shown that the electrical noise in the current that is measured in the patch electrode is given by $4kT/R_{leak}$, where 4kT is a constant. Thus, as R_{leak} increases (i.e. as the seal improves during suction) the current noise gets smaller. This can be seen in Figure 6.1 C, where the noise level falls dramatically following suction. Examples of the noise in the measured current before and after sealing are shown expanded in the lower part of C (see also Hamill et al., 1981).

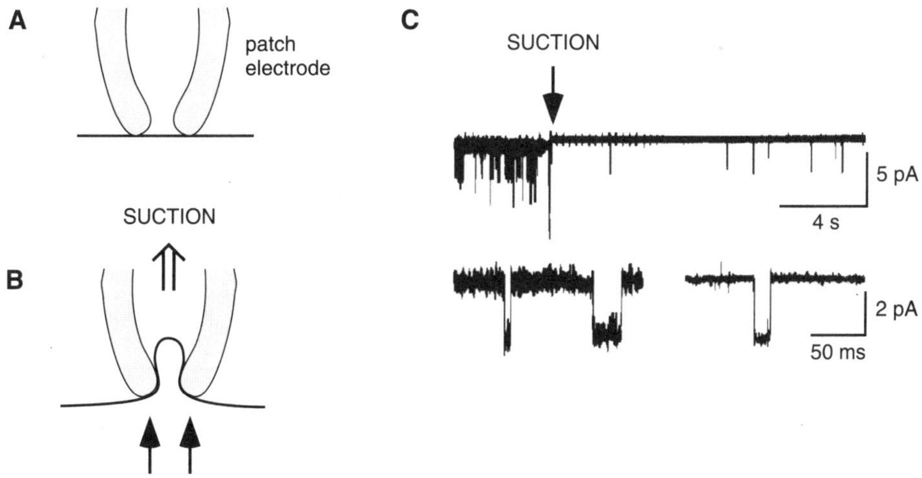

Figure 6.1 Getting a gigaseal. (Modified from Hamill et al., 1981.)

In practice, and with luck, seal resistances in excess of 1 GigaOhm ($G\Omega = 10^9\Omega$) can be obtained on many kinds of cells; this is the so-called 'gigaseal'. With a gigaseal, noise levels can be so low that the current flowing through individual ion channels in the membrane patch can easily be resolved. For example, the traces in Figure 6.1 C show current steps due to the opening of single acetylcholine-gated channels (downward deflections).

The gigaseal is usually so stable that a number of further manipulations are often possible following its formation (Hamill et al., 1981). These are summarized in Figure 6.2. The configuration described above is the 'cell attached' mode. It is useful for studying single channels in intact membrane without disturbing the interior of the cell. By applying a brief, strong pulse of suction to the back of the patch electrode, it is usually possible to rupture the patch without breaking the seal. This is 'whole-cell recording', in which the interior of the patch electrode is contiguous with the interior of the cell. This recording configuration is commonly used to replace 'sharp' intracellular electrodes, since it offers excellent access to the cell interior, low leakage, and good stability. On the other hand, with this configuration it is possible to wash biologically important factors out of the cell (see below).

If, following attainment of the whole-cell mode, the patch electrode is slowly backed away from the cell, the 'outside-out patch' configuration results. This works because cell membranes are fluid; as the membrane is stretched out, it eventually pinches off to form a new patch of membrane across the opening at the electrode tip, leaving the outer surface of the membrane facing the external solution (hence: 'outside [surface facing] out' patch). Like the cell-attached mode, outside-out patches permit study of single-channel currents, with the advantage that both sides of the membrane are now accessible to artificial solutions. Moreover, the extracellular side is especially accessible, since it faces the bath. This permits rapid exchange of external solutions (see Chapter 11).

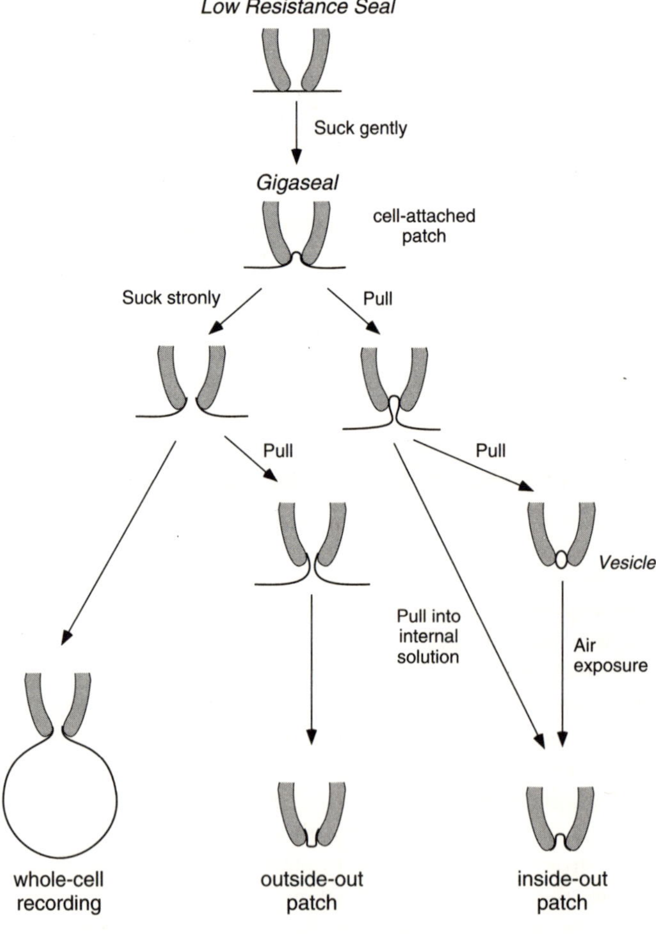

Figure 6.2 Patch-clamp recording configurations. (Modified from Hamill et al., 1981.)

Finally, the 'inside-out patch' configuration is the reverse of the outside-out patch. To form an inside-out patch, a cell-attached patch is simply pulled off the cell. Often, however, the membrane pinches off when this is done, forming a vesicle at the tip of the electrode (Figure 6.2). The unwanted external side of the vesicle can be removed by briefly lifting the patch out of the bath; with luck, only the outer leaflet of membrane breaks. Alternatively, the cell-attached patch can be pulled off into a low Ca^{2+} solution (i.e. one that resembles internal solution). The main advantage of the inside-out patch is the ease of changing solutions that bathe the cytoplasmic face of the membrane. Note that the choice of solution to put in the patch electrode depends on the recording configuration. For whole-cell and outside-out patches, the electrode contains 'internal' solution. For cell-attached and inside-out, it contains 'external' solution.

USES OF THE PATCH-CLAMP TECHNIQUE

This section gives a very brief overview of some of the applications of patch-clamping, in order to convey a sense of the versatility of the technique.

1. Whole-cell recording of ionic currents

Whole-cell recording of ionic current is the patch-clamp equivalent of experiments with 'sharp' intracellular electrodes. There are two major advantages of the whole-cell approach: (a) much reduced noise in the recordings, owing to the high resistance glass-membrane seal, and (b) much better voltage- and solution-clamping of the interior of the cell because of the relatively large tip diameter (and thus low resistance) of the patch electrode. Better voltage-clamping means that fast currents are recorded more accurately and the cell's membrane potential is more tightly controlled by the electronics of the amplifier. Better solution clamping means that the contents of the patch electrode diffuse more rapidly and completely into the cell. The latter can be a disadvantage if important soluble molecules are washed out of the cell. However, a variant of whole-cell recording, called perforated patch-clamp, overcomes this problem. In this technique, a pore-forming agent (e.g. nystatin) is added to the internal solution in the patch electrode (Horn & Marty, 1988). Following seal formation, rather than rupturing the patch under the electrode tip, one simply waits for the nystatin pores to insert into the patch. The pores are too small to allow larger or highly charged molecules to pass, but electrical access is mediated by permeant small monovalent ions.

Traditionally, patch-clamping was restricted to cultured or acutely dissociated cells, because they lack much of the connective tissue and other impedimenta to the approach of a patch electrode (see Chapter 2). However, in 1989 it was first shown that acute brain slices can also be patch-clamped with surprising ease (Blanton et al., 1989; Edwards et al., 1989). This application of the technique, sometimes called the 'patch-slice' technique, has increased explosively since then. A recent improvement of the patch-slice technique uses infrared optics to visualize neurons deep within slices, allowing very precise placement of patch electrodes onto dendrites, or even axons (Stuart et al., 1993).

2. Second messengers

Creative use of patch-clamping has answered a number of questions about second messenger cascades in cells. An example is a recent study of the modulation of calcium channels by muscarinic receptors in rat sympathetic neurons (Mathie et al., 1992). In this experiment, a cell-attached patch was formed and calcium channels in the patch were activated by voltage steps. A muscarinic agonist was then added to the bath solution, resulting in a suppression of the calcium channel activity in the patch. Since the muscarinic agonist could not leak under the gigaseal to directly modulate the calcium channels, it must have been activating muscarinic receptors elsewhere on the cell. These receptors must then have released a second messenger that could diffuse freely through the cytoplasm of the cell, thereby modulating the calcium channels in the otherwise isolated patch of membrane. Thus, this experiment provided direct evidence for freely diffusible cytoplasmic second messengers. In other types of experiments, putative second messengers can be tested by directly applying them to the cytoplasmic face of an inside-out patch and looking for channel modulation.

3. Single-cell molecular biology

This third application of patch-clamping has been attracting attention only within the past few years. The goal is to use molecular biological techniques (e.g. PCR; see Chapter 30) to provide a profile of the kinds of messenger RNA (mRNA) that are found in a single cell, because this indicates what proteins might be expressed in that cell. In one type of experiment, molecular biology reagents are mixed with the solution in the patch electrode, which is then used to obtain a whole-cell recording from the cell of interest. The reagents diffuse into the cell, which is now a tiny reaction vessel for, typically, the synthesis of complementary DNA (cDNA) from the cell's mRNA. The contents of the cell are then aspirated into the patch electrode, expelled into a plastic tube, and further stages of the reaction carried out. Since the gigaseal is so stable, it is often possible to aspirate most of the cell's contents without breaking the seal. This ensures that contamination from the cytoplasms of other cells is avoided. A further attraction of this technique is that the cell can be characterized electrophysiologically during the time the cDNA is being synthesized with the cell held in the whole-cell mode. Thus, the single patch electrode allows both the molecular and functional profile of the cell to be determined (Sucher & Deitcher, 1995).

REFERENCES AND FURTHER READING

Blanton, M. G., Lo Turco, J. J. and Kriegstein, A. R. (1989) Whole cell recording from neurons in slices of reptilian and mammalian cerebral cortex. *J. Neurosci. Meth.* **30**, 203–210.

Edwards, F. A., Konnerth, A., Sakmann, B. and Takahashi, T. (1989) A thin slice preparation for patch-clamp recordings from neurons of the mammalian central nervous system. *Pflügers Arch.* **414**, 600–612.

Hamill, O. P., Marty, A., Neher, E., Sakmann, B. and Sigworth, F. J. (1981) Improved patch-clamp techniques for high-resolution current recording from cells and cell-free membranes. *Pflügers Arch.* **391**, 85–100.

Horn, R. and Marty, A. (1988). Muscarinic activation of ionic currents measured by a new whole-cell recording method. *J. Gen. Physiol.* **92**, 145–159.

Mathie, A., Bernheim, L. and Hille, B. (1992). Inhibition of N- and L-type calcium channels by muscarine receptor activation in rat sympathetic neurons. *Neuron* **8**, 907–914.

Neher, E. (1992). Ion channels for communication between and within cells. *Science* **256**, 498–502. (Nobel Prize Lecture).

Neher, E. and Sakmann, B. (1992). The patch-clamp technique. *Sci. Am.* **266**, 28–35.

Sakmann, B. (1992). Elementary steps in synaptic transmission revealed by currents through single ion channels. *Science* **256**, 503–512. (Nobel Prize Lecture).

Sakmann, B. and Neher, E. (eds) (1995). *Single-channel Recording*, (2nd edn). New York: Plenum Press.

Stuart G. J., Dodt H. -U. and Sakmann, B. (1993). Patch-clamp recordings from the soma and dendrites of neurons in brain slices using infrared video microscopy. *Pflügers Arch.* **423**, 511–518.

Sucher, N. J. and Deitcher, D. L. (1995) PCR and patch-clamp analysis of single neurons. *Neuron* **14**, 1095–1100.

CHAPTER 7

ANALYSIS OF SINGLE CHANNEL RECORDINGS

D. Laver
Division of Neuroscience, John Curtin School of Medical Research, Australian National University

INTRODUCTION

This chapter covers the basic principles of data acquisition and analysis. It is assumed that the reader is already familiar with techniques for measuring and recording ion channel signals. Data acquisition (digitizing) is the process by which an analogue signal, usually replayed from a recording device, is transformed into a series of data points which describes the time-course of that signal. Signal analysis involves the derivation of the signal statistics which can be used to test hypotheses about how channels work or their role in cell function. Analysis of ion channel signals spans a wide range of complexity. This chapter deals with some of the most basic and commonly used methods for describing ion channel signals. (For a more detailed account of data acquisition and analysis refer to Sakmann and Neher (1983).)

SIGNAL ACQUISITION: NOISE AND FILTERING, DEATH AND TAXES

While some may argue that taxes and even death are avoidable, no one escapes noise and filtering. Background electrical noise arises from a variety of sources (e.g. thermal fluctuations and the discrete nature of electronic charge) and superimposes random fluctuations which tend to obscure the ion channel signal. Since noise is random its properties must be described by statistical parameters. For example, the size of the noise is measured by the root-mean-square of the current, I_{RMS}, which is the same as the standard deviation of the signal. Another important statistic is the frequency which shows the relative contribution of different frequency components of the noise. The nature of electrical noise in biological membranes is such that the higher frequency components are larger than the lower frequencies.

Filtering is the means of distorting signals with a view to revealing specific signal characteristics. Filtering can be done by passing the signal through an electric circuit (analog filter) or by transforming the digitized signal data using a computer (digital filter). Filters are classified according to their attenuating affect of different parts of the signal spectrum. For example high-pass filters attenuate lower frequencies, low-pass filters attenuate higher frequencies and band-pass filters retain signals over a specific frequency range. Low-pass filters are frequently used to reduce the background noise on ion channel signals (Figure 7.1). Two important characteristics of low-pass filters are the corner (or cut-off) frequency and the frequency dependence of signal attenuation

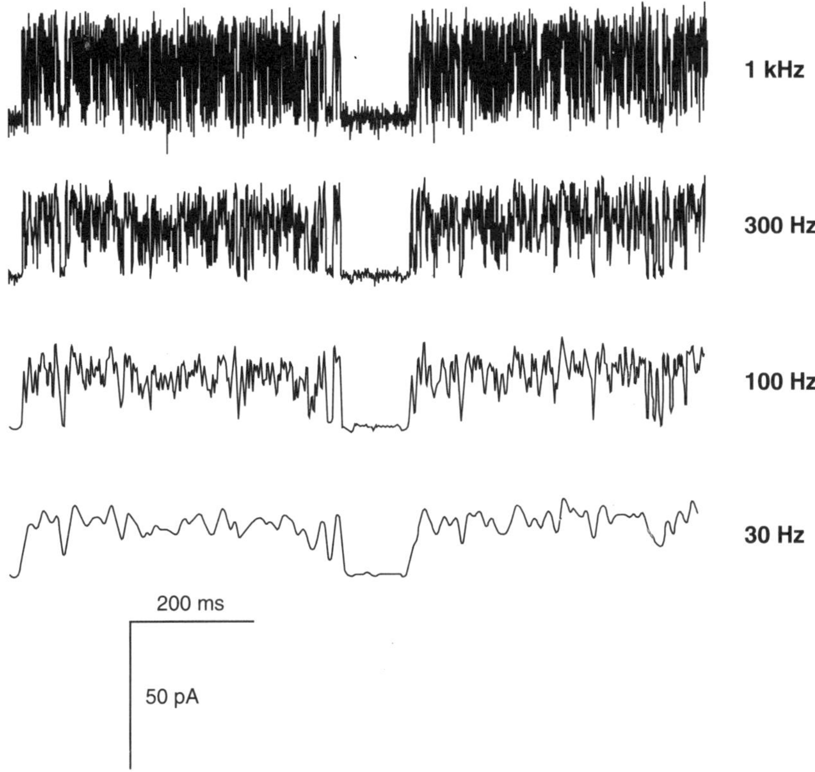

Figure 7.1 The current through a cation channel which has been filtered using a Bessel filter with the corner frequencies listed on the right.

(frequency response, or roll-off; see Figure 7.2 A). The effect of filter cut-off frequency on ion channel signal can be seen in Figure 7.1. As the cut-off frequency is lowered the magnitude of the noise is reduced. However, associated with this distortion of the signal is a blurring of the shorter channel openings and closings. Two examples of low-pass filters with different role-off characteristics are the Bessel and Butterworth filters. Butterworth filters have a flatter response at low frequencies and a sharper cut-off than the Bessel filter. However, Bessel filters are preferred for channel recordings because they produce a more desirable distortion of current jumps (time response: see Figure 7.2 B). Another characteristic of the filter role-off is determined by the number of poles in the filter (also referred to as the order of the filter). The effect of an N-pole filter is the same as filtering the signal N times through a single pole filter. Consequently, the frequency response of a multipole filter falls off more rapidly than a single pole filter (Figure 7.3). Bessel filters with four to eight poles are commonly used for ion channel signals. It must be remembered than even when filtering is not explicitly used, filtering in some form is unavoidable. All signal measurements incur distortion because of the finite response time of the measuring apparatus to abrupt changes in signal amplitude.

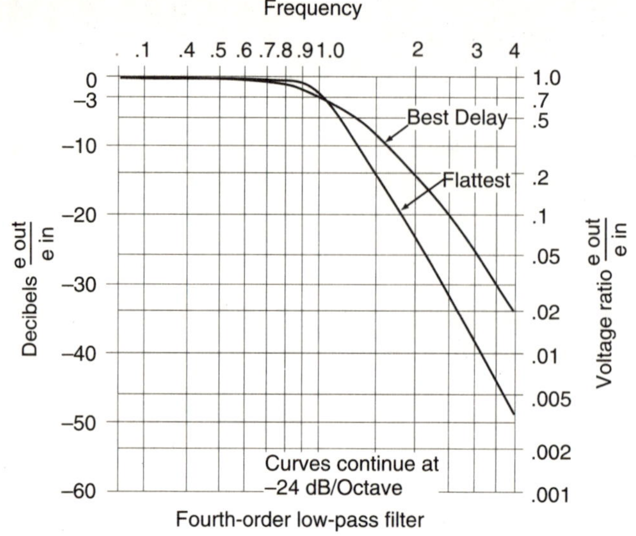

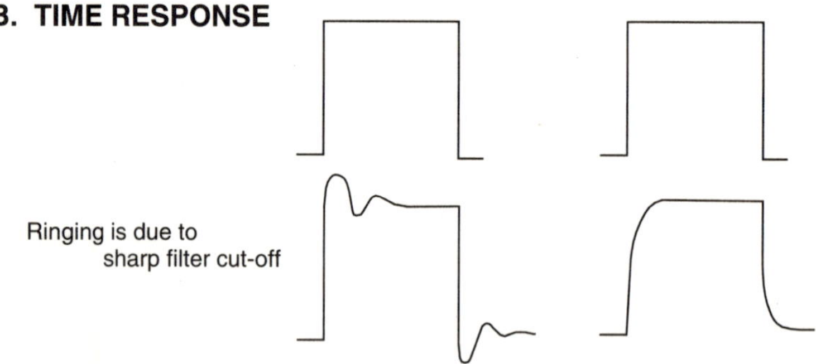

Figure 7.2 Comparison of Bessel and Butterworth low-pass filters. A: The frequency response of Butterworth (labelled Flattest) and Bessel (labelled Best Delay) filters where the corner frequency is 1.0 Hz. B: A schematic diagram showing the effect of Butterworth (left) and Bessel (right) filters on a square-wave signal showing the time response of these filters.

Analog signals are converted to a series of data points by sampling the signal at discrete, equally spaced time intervals with an analog-to-digital converter (ADC) under the control of a computer. ADCs commonly used to sample ion channel recordings convert the signal into a 12-bit binary code. This means that the data can have any one of 2^{12}–1 (4095) values. If the ADC can sample over the range ±5 V then signal amplitudes measurements can only take on values which are multiples of ~2.5 mV. Thus, unlike the analog signal, the digitized signal is quantized both in time and amplitude and is therefore only an approximation to the original signal. In order to produce an accurate representation of the original signal it is necessary to sample it in the correct way. The

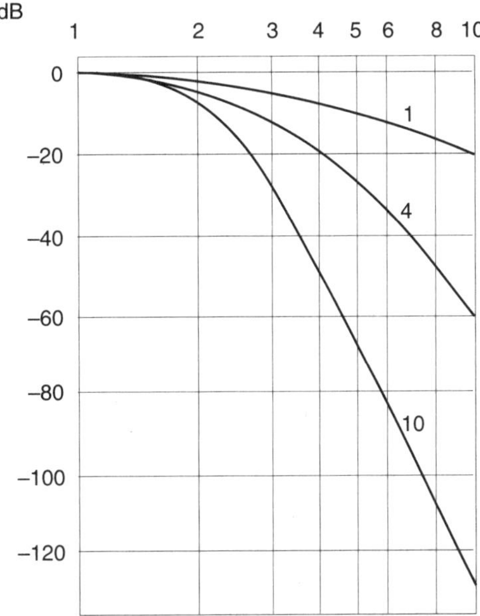

Figure 7.3 The frequency response of Bessel filters with 1, 4 and 10 poles. The vertical scale in dB is a log scale where –20 corresponds to a 10-fold attenuation.

total amplitude of the signal should be an appreciable fraction of the range of the ADC so that quantization of the signal only introduces small errors. A more subtle digitizing artifact is 'aliasing' which occurs when the signal is sampled too infrequently (Figure 7.4). As shown in Figure 7.4, a high frequency sinusoidal wave can appear as a low frequency signal. The Nyquist frequency is the minimum sampling rate which avoids aliasing and this is equal to twice the signal frequency. Aliasing can increase the background noise level and cause the omission of short current levels in channel recordings. Hence the sampling frequency must exceed double that of the highest frequency component in the signal. However, because filters have a gradual roll-off the highest frequency in the signal is not defined. In practice, when using 4–8 pole Bessel filters a sampling rate of five times the filter corner frequency is used. Higher sampling rates improve the signal rendition but at the same time consume more computer memory.

SIGNAL ANALYSIS

This section is devoted to examining the most commonly used methods for describing the properties of ion channel signals. These signals generally appear as a series of stochastic current jumps between stationary levels (see, e.g. Figure 7.1). For most channel types the jumps are mainly between the closed state and a maximum level (the unitary current). Two related statistical parameters which give an overall measure of the channel

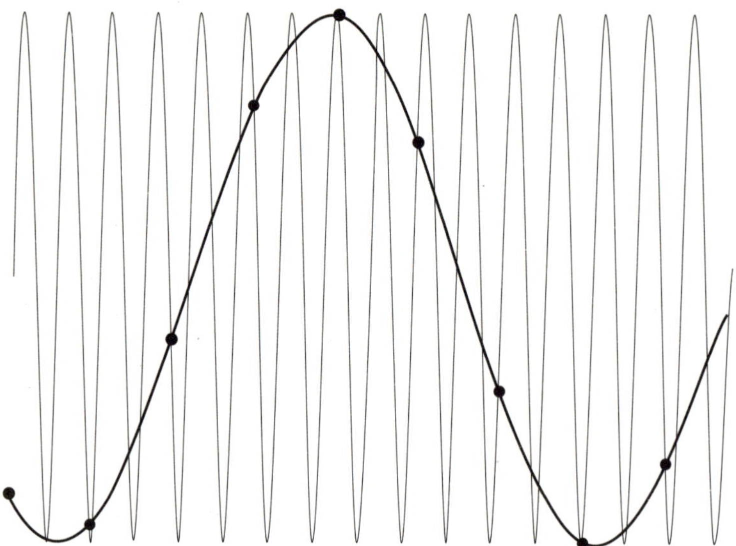

Figure 7.4 The effect of sampling a signal too slowly. The original sine-wave signal is shown by the fine line. This data is sampled at equally spaced time intervals shown by the dark circles. Joining the dots reveals a sine wave with much lower frequency than the original signal. This low frequency signal is an artifact of the sampling method and is called an alias signal. The minimum sample rate needed to avoid aliasing is double the frequency of the original signal (Nyquist frequency). In practice it is advisable to sample at ~5 times the signal frequency to get a reasonable rendition of the signal.

activity are the open probability, P_o, and the normalized, mean current, I′. P_o is the proportion of the time the channel is in a conducting state, and is calculated from the ratio of number of data points in conducting levels and the total number of points in the record (assuming equally spaced data). I' is equal to the mean of every current amplitude divided by the unitary current. These parameters are useful because they measure the 'openness' of the channel and give an indication of what the current would be through the whole channel population. It turns out that for ion channels with only one open conductance level P_o and I′ are the same.

More detailed analysis of channel signals identifies the various current levels the channel can produce and the time spent at each level before the conductance changes to a new value. The most popular method for visualizing current levels from a channel signal is the all-points amplitude histogram. This is a histogram of all data points grouped according to their amplitude (Figure 7.5). Peaks in the histogram correspond to sustained levels in the record. The width of the peaks is proportional to the size of the background noise and the area under each peak is proportional to the total time spent at that level.

Frequency histograms of open and closed duration provide clues to the mechanisms responsible for opening and closing ion channels. Ion channel signals, in general, have many more short open and closed durations than long ones, and the frequency

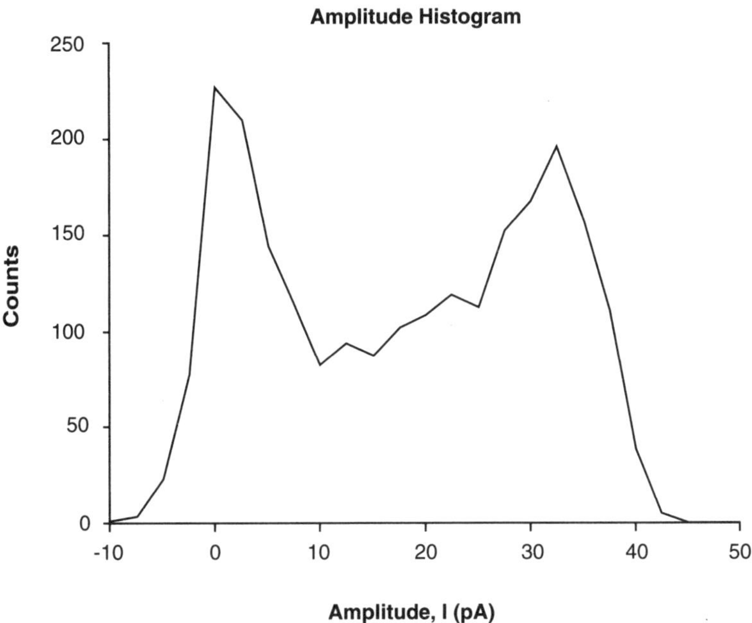

Figure 7.5 The amplitude histogram of all data points in the record shown in Figure 7.1 (1 kHz). The peaks at ~0 and 30 pA correspond to the closed and open current levels in the record.

distribution of these events can be described by the sum of decaying exponential functions with various time constants. There are several graphical methods for displaying these histograms, four of which are shown in Figure 7.6. The most useful type of plot, as shown in Figure 7.6 C and D, is where the data are grouped into bins that are equally spaced on a log scale (log-binned; the bin width increases along the time axis). In the log-binned scale the broadening of the bins at longer times increases the number of counts in each bin which tend to counter the exponential decline seen in uniformly binned distributions (Figure 7.6 A and B). On the log-binned histograms the exponentially distributed frequencies on the linear scale (Figure 7.6 A) form peaked distributions where each peak corresponds to an exponential component of the distribution. When the square root of the frequency is log-binned (Figure 7.6 D) the statistical scatter on the data is uniform across the entire distribution. This is a useful property for when the data is to be fitted with theoretical functions during subsequent data analysis.

REFERENCES AND FURTHER READING

Sakmann, B. and Neher, E. (eds) (1983) *Single-Channel Recording*. New York: Plenum Press.

Sigworth, F. J. and Sine, S. M. (1987) Data transformations for improved display and fitting of single-channel dwell time histograms. *Biophys. J.* **52**, 1047–1054.

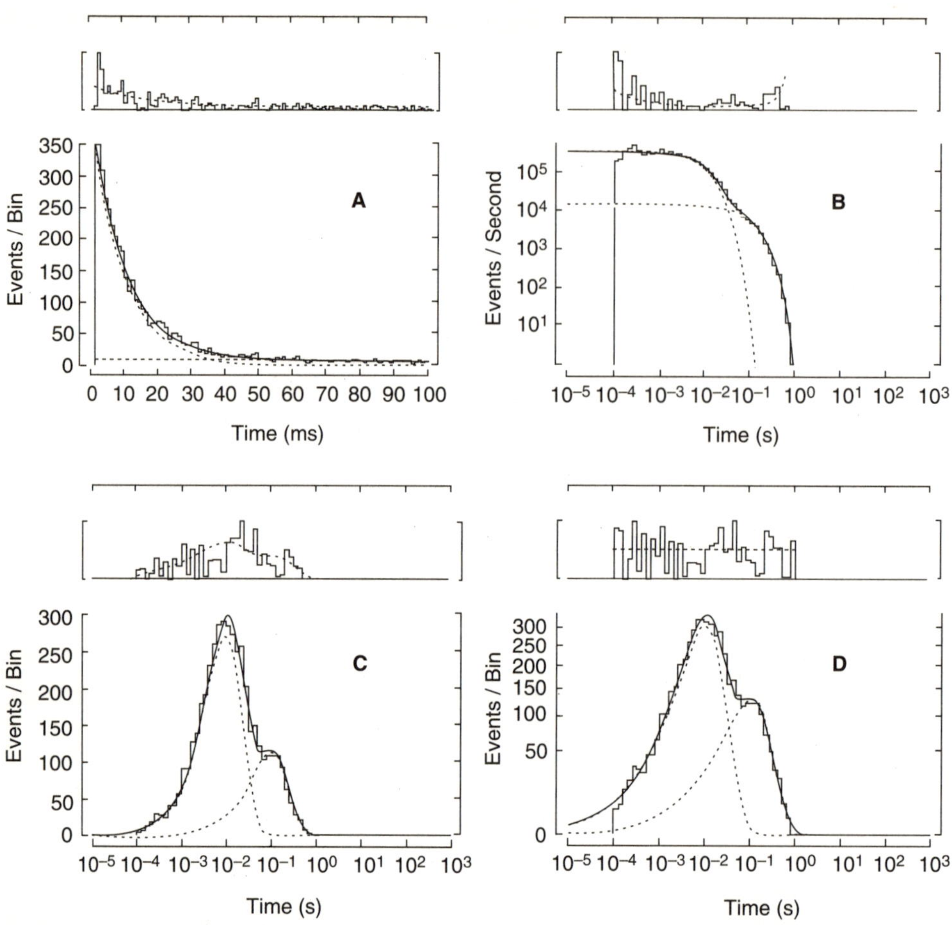

Figure 7.6 Four methods for plotting durations of channel open and closed events. (The same data is used for each method.) The statistical scatter of the data resulting from a limited sample size (shown above each graph) was determined from the differences between the event frequencies and a smooth curve fitted to the data. A: The event frequency (events/bin) are shown for uniform bin widths which shows a typical, exponential, frequency distribution seen for ion channels' openings and closures. B: The bins have uniform width but the data is plotted on a log scale. Each exponential component of the data (there are two) appears as a shoulder on the distribution. The component with the shorter exponential time constant appears as the left shoulder. Note that the exponential component with the longer time constant is not obvious in A. C: Here the data are grouped into bins that increase in width in proportion with time and so are equally spaced on a log scale. The wider bins tend to collect more counts than the narrow bins and so get more heavily weighted in the distribution. Using this format the exponential components form peaked distributions where each peak position coincides with the time-constant of that exponential component. D: The data have been grouped as in C except that the square root of the frequency is plotted here. Note that the statistical scatter on the data is uniform across the entire distribution. (Redrawn from Figure 1 in Sigworth & Sine (1987).)

SECTION 3

HUMAN ELECTROPHYSIOLOGY

CHAPTER 8

RECORDING ELECTROMYOGRAPHIC ACTIVITY (EMG) FROM SINGLE MOTOR UNITS IN HUMAN SUBJECTS

V. G. Macefield
Prince of Wales Medical Research Institute, Randwick, Sydney

INTRODUCTION

Recording electromyographic activity (EMG) in human subjects is rather straightforward: apply one Ag–AgCl surface electrode over the muscle belly, another closer to the tendon, amplify and filter the signal (100–1000×, band-pass 0.01–1 kHz) and get the subject to do his or her thing. Modern EMG amplifiers are 'differential', in that the two inputs (the 'active' electrode over the muscle belly and the 'reference' electrode near the tendon) are electronically subtracted from each other, resulting in high 'common-mode' rejection (i.e. background noise is largely eliminated). The inputs also have high impedances which further improves signal quality. An important consideration for human studies is that the amplifiers need to be electrically isolated (the newer models use optoelectronic isolation) from the mains power supply to avoid the risk of electric shock. For most studies it is enough to record from a large sample of the muscle, that is, record multi-unit EMG, and quantify the response by one of several means. Typical methods of quantification include integrating the signal with a 'leaky' integrator (e.g. a resistor-capacitor circuit with a time constant of 100 ms), generating a 'moving average', root-mean-square (RMS) processing or some other form of electronic gymnastics which converts the action potentials generated by many motor units into a nice smooth curve from which one can measure amplitude and/or area. This is fine if the signal is not contaminated with activity from other nearby muscles ('cross-talk'), and if care is taken not to over-interpret the EMG signal in terms of 'recruitment' and 'rate-coding' within the motoneuron pool. Cross-talk can be a very real problem with surface electrodes and is minimized by accurate location of the electrodes with respect to the muscle's anatomy, minimizing the inter-electrode distance, and attempting to limit the contraction to the muscle under study. As to the issues of recruitment and rate-coding, these can only be studied effectively if single motor units (i.e. the muscle fibers supplied by individual motoneurons) can be adequately isolated or discriminated within a recording. For a critical assessment of the theoretical constraints for detecting motor unit activity with surface electrodes, see the paper by Fuglevand et al. (1992).

METHODS OF RECORDING FROM SINGLE MOTOR UNITS

Recording from many active motor units during a contraction has been likened to listening to the overall level of conversation in a people-packed room with an omnidirectional microphone; to listen in on one person's voice a close focal microphone is required. As an interesting aside, the sounds associated with a muscle contraction can be recorded using piezoelectric microphones attached to the overlying skin, a technique known as acoustic myography (Barry et al., 1985). While it is possible to record electromyographic activity from single motor units using surface electrodes when a contraction is weak (i.e. if only one person is speaking in our analogy), if it is too strong the recruitment of additional units, and the superimposition of action potentials from synchronously active motor units, will prevent adequate isolation of the motor unit's discharge. Indeed, when stronger contractions are required it is better to record intramuscularly using hook-wire electrodes. These can be fashioned by inserting two lengths of insulated 0.1 mm diameter nichrome wire through a 23-gauge hypodermic needle and exposing 1–2 mm at the ends by scraping off the insulation. These ends are then bent back against the shaft of the needle, forming two barbs. After sterilization the entire assembly can be inserted through the skin into a muscle belly and the needle withdrawn, leaving the wires behind. Slight adjustments in the intramuscular location of one or both ends can be made by gently pulling on the wires, but as a rule these electrodes, once inserted, cannot be relocated. Needless to say, this is one of their particular advantages: their position is stable enough, and the wires flexible enough, to allow gross movements without fear of loosing the recording site. Depending on the intensity of the contraction these electrodes generally pick up the electrical signals of a few or many motor units. A more selective recording can be obtained by leaving the insulation in place, such that the recording surfaces are limited to the cut ends of the wires. This allows single motor units to be followed during even higher contraction levels. There are other variations: removing a small area of insulation from the middle of a length of wire (rather than from the ends) and running two of these wires through the skin and out the other side, leaving the wires (and exposed recording surfaces) within the muscle belly. Monopolar recording electrodes (in which the reference electrode is located outside the muscle) offer the advantage that they can be moved within the muscle if necessary, as well as providing a well-defined, small recording surface.

RECORDING FROM SINGLE MOTOR UNITS DURING MAXIMAL CONTRACTIONS

For recording single motor units during maximal voluntary contractions, the only possible way so far has been to use insulated tungsten microelectrodes (the same as those used for human microneurography; see Chapter 9). These have a naturally high impedance, limiting the recording field to a very small area around the tip. They also allow the recording site to be adjusted which, because of the high forces generated during a maximal voluntary contraction (which tend to push out the microelectrode) is somewhat hard to avoid. Indeed, one tends to sample short trains of single motor units during a maximal contraction by manually adjusting the electrode depth throughout the contraction; in this way a series of spike trains are recorded at different times which, when averaged with other trains recorded during subsequent contractions, allows the construction of a time-varying picture of the changes in firing rate during a maximal voluntary contraction. Nevertheless, it is possible to follow a single unit throughout much of the contraction by manually applying a force which counteracts that tending to expel the microelectrode. Table 8.1 shows the maximal firing rates (mean ± SD) recorded from a range of human muscles, arranged in descending order.

Table 8.1 Mean (± SD) firing rates recorded from single motor units in different muscles during maximal voluntary contractions in human subjects.

Muscle	*Maximal rate (Hz)*	*Reference*
biceps brachii	31.1 ± 10.1	Bellemare et al. (1983)
adductor pollicis	29.9 ± 8.6	Bellemare et al. (1983)
tibialis anterior	28.2 ± 9.9	Bigland-Ritchie et al. (1992)
extensor hallucis longus	19.5 ± 8.0	Fuglevand et al. (1994)
soleus	10.7 ± 2.9	Bellemare et al. (1983)

The above numbers are important in that they demonstrate the low firing rates of human motoneurons. If recorded mean firing rates are appreciably higher than these then something is wrong, for instance, the recording may be from two units of similar amplitude and shape. In addition to this low upper limit of firing, the minimal firing rates attainable during a weak voluntary contraction are low—most units fire at about 8 Hz following recruitment. Consequently, the range of discharge frequencies one can expect to record from single motor units during voluntary contractions is quite narrow (Macefield et al., 1993). As a physiological aside, the maximal firing rates that motor units in a particular muscle are capable of generating fit the known histochemical profiles and physiology of different muscles. Microstimulation of single motor axons in the median and peroneal nerves (Thomas et al., 1991; Macefield et al., 1996) has shown that, because they have faster contractile speeds, motor units in the short muscles acting on the thumb require higher frequencies of stimulation to generate half their peak force than do motor units in the long muscles acting on the toes, and these fundamental differences in contractile properties are reflected in their firing rates during maximal voluntary contractions. One final methodological issue needs to be addressed: the state of the muscle when recording from single motor units. If the subject has performed many contractions neuromuscular fatigue may set in, resulting in motor unit firing rates that are different from those recorded in 'fresh' muscle. This is particularly pronounced during maximal voluntary contractions when maximal firing rates fall progressively during the contraction and may take some time to recover (Bigland-Ritchie et al., 1983, 1992; Fuglevand et al., 1994).

RECORDING FROM SINGLE MUSCLE FIBERS

In addition to recording from individual motor units, one can also record from a single fiber within a motor unit using a special monopolar needle electrode. Here, the recording surface is limited to a small window (25 μm diameter) on the side of the needle. Single-fiber EMG is used diagnostically in many clinical neurophysiology laboratories for assessing muscle fiber density in patients (Sanders & Stålberg, 1996). This allows examination of muscle fiber topography within a motor unit without removing a sample of tissue through a large-diameter needle (needle biopsy). Furthermore, by recording from pairs of muscle fibers during a voluntary contraction one can measure the normal variations in delay between a motor axon and the muscle fibers it supplies, the so-called 'neuromuscular jitter'. Again, this can be used clinically to assess abnormalities in neuromuscular function (Sanders & Stålberg, 1996).

RECENT PAPER

Cross-correlations between single human motor units during weak voluntary contractions is a fairly hot topic. This approach determines how time-locked two simultaneously active motor units can be. The results provided in the study by Bremner et al. (1991) are quite dramatic: they found that 88% of the motor units recorded from one muscle or from a coactivated muscle appeared to share a common synaptic drive. The interpretation of this data is that the majority of spinal motoneurons active during a particular contraction may receive excitatory synaptic drives from the same upper motoneuron.

REFERENCES AND FURTHER READING

Barry, D. T., Geiringer, S. R. and Ball, R. D. (1985) Acoustic myography: a noninvasive monitor of motor unit fatigue. *Muscle and Nerve* **8**, 189–194.

Bellemare, F., Woods, J. J., Johansson, R. and Bigland-Ritchie, B. (1983) Motor-unit discharge rates in maximal voluntary contractions of three human muscles. *J. Neurophysiol.* **50**, 1380–1392.

Bigland-Ritchie, B., Furbush, F. F., Gandevia, S. C. and Thomas, C. K. (1992) Voluntary discharge frequencies of human motoneurons at different muscle lengths. *Muscle and Nerve* **15**, 130–137.

Bigland-Ritchie, B., Johansson, R. S., Lippold, O. C. J., Smith, S. and Woods, J. J. (1983) Changes in motoneuron firing rates during sustained maximal voluntary contractions. *J. Physiol.* **340**, 335–346.

Bremner, F. D., Baker, J. R. and Stephens, J. A. (1991) Correlation between the discharges of motor units recorded from the same and from different finger muscles in man. *J. Physiol.* **432**, 355–380.

Fuglevand, A. J., Macefield, V. G., Howell, J. N. and Bigland-Ritchie, B. (1994) Adaptation of motor unit discharge rate and variability during sustained maximal voluntary contractions. *Soc. Neurosci. Abstr.* **20**, 1759.

Fuglevand, A. J., Winter, D. A., Patla, A. E. and Stashuk, D. (1992) Detection of motor unit action potentials with surface electrodes: influence of size and spacing. *Biol. Cybernetics* **67**, 143–153.

Macefield, V. G., Fuglevand, A. J. and Bigland-Ritchie, B. (1996) Contractile properties of single motor units in human toe extensors assessed by intraneural motor-axon stimulation. *J. Neurophysiol.* **75**, 2509–2519.

Macefield, V. G., Gandevia, S. C., Bigland-Ritchie, B., Gorman, R. B. and Burke, D. (1993) The firing rates of human motoneurons voluntarily activated in the absence of muscle afferent feedback. *J. Physiol.* **471**, 429–443.

Thomas, C. K., Bigland-Ritchie, B. and Johansson, R. S. (1991) Force-frequency relationships of human thenar motor units. *J. Neurophysiol.* **65**, 1509–1516.

Sanders, D. B. and Stålberg E. V. (1996) Single-fiber electromyograhy. *Muscle and Nerve* **19**, 1069–1083.

Thomas, C. K., Bigland-Ritchie, B. and Johansson, R. S. (1991) Force-frequency relationships of human thenar motor units. *J. Neurophysiol.* **65**, 1509–1516.

CHAPTER 9

INTRANEURAL RECORDING FROM SINGLE NERVE FIBERS IN AWAKE HUMAN SUBJECTS

V. G. Macefield
Prince of Wales Medical Research Institute, Randwick, Sydney

INTRODUCTION

Just think: to be able to record from single nerve fibers in human subjects and gain the same kind of information as that obtained from splitting nerve fibers under a microscope in an anaesthetized, and often pharmacologically paralysed, experimental animal. Indeed, it is possible to make beautiful single unit recordings from peripheral nerves in awake human subjects, through the technique known as *microneurography*. This was developed in the late 1960s by Karl-Erik Hagbarth and Åke Vallbo at the University Hospital in Uppsala, Sweden. It involves inserting a tungsten microelectrode through the skin and manually directing it into an accessible peripheral nerve: the median or ulnar nerve below the biceps muscles, the radial nerve below the triceps muscles, the median nerve between the flexor tendons at the wrist, or the peroneal nerve as it courses around the head of the fibula at the knee. The first recordings were made from single primary afferent axons supplying mechanoreceptors in the muscles and skin (Hagbarth & Vallbo, 1969; Knibestol & Vallbo, 1970) but this has since been extended to joint receptors in the fingers (Edin, 1990; Macefield et al., 1990) and single motor axons supplying skeletal muscles (Macefield et al., 1993). Single-unit recordings have also been made from sensory nerve fibers innervating the teeth, via microelectrodes inserted into the inferior alveolar nerve through the open mouth (Trulsson et al., 1992). Other cranial nerves studied include the facial nerve, and the supraorbital and infraorbital branches of the trigeminal nerve (Johansson et al., 1988; Nordin, 1990).

LOCATING THE NERVE AND ENTERING A FASCICLE

Certain nerves can be found by palpation and knowledge of specific anatomical landmarks, but generally an external stimulating electrode is used to locate the nerve by delivering weak electrical pulses (<10 mA, 0.2 ms, 1 Hz) through the skin. The site of lowest threshold for evoking pins-and-needles (paraesthesiae) in the cutaneous distribution of the nerve, and/or twitches in the innervated muscles, is then marked as the site for inserting the microelectrode. The electrodes used for percutaneous recordings from human nerves are made from the stiff metal tungsten, and typically have a shaft diameter of 0.2 millimeters and a tapered point that is sharpened mechanically or electrolytically to a diameter of a few microns. The entire electrode is insulated by repeated dipping in lacquer, or by vacuum-deposition of parylene. While some laboratories produce their own, commercial sources of tungsten microelectrodes include Fredrick Haer Co. and World Precision Instruments in the US. After inserting the microelectrode through the skin for a few mm, weak negative pulses are applied through the electrode which is then advanced manually until the nerve sheath and then a nerve fascicle is entered, heralded by the subject's reports of stronger evoked sensations and/or observed twitches. An uninsulated subdermal electrode inserted nearby serves as the reference. Once the stimulating current required for generating sensations/twitches is below 0.025 mA, the electrode is switched to a preamplifer-amplifier system (total gain 10 000x, band-pass 0.3–5 kHz). Stroking the skin in the innervation territory of the fascicle, or tapping over relevant tendons or muscles, will then elicit multi-unit afferent activity which can be monitored visually and audiometrically. It is then just a matter of making slight (very slight) manual adjustments of the electrode's position within the fascicle until large spikes (action potentials) appear out of the multi-unit evoked discharge. Because micromanipulators cannot be used (ongoing respiratory and pulse fluctuations would cause relative motion between a mechanically fixed electrode and an intraneural recording site), the only sources of stability are the skin and underlying tissues and the epineural and perifasicular nerve sheaths. In some subjects with low body fat the nerve of interest may be so superficial that there is insufficient tissue to keep it in place, whereas in subjects with a higher per cent of body fat there may be so much adipose tissue that the nerve is difficult to find or the electrode is not long enough. A general limitation, though, is that the subject must sit very still for the duration of the experiment—slight adjustments in body position can result in loss of a unitary recording site. Experiments usually last 3–4 hours. Nevertheless, stable recordings from single nerve fibers can often be held for up to an hour if necessary, although rarely are such long recordings required. And a particular advantage of doing experiments in awake human subjects is that they can be asked to perform specific tasks such as contracting muscles to a target force (Macefield et al., 1991), or manipulating a gripped object (Westling & Johansson, 1987).

GENERAL FEATURES OF MICROELECTRODE RECORDINGS FROM HUMAN NERVES

The majority of recordings from large-diameter nerve fibers (axons) have positive-going spikes, which reflect impalement of the myelin sheath by the electrode tip in the internode region (Vallbo, 1976; Inglis et al., 1997). Negative-going spikes, indicating proximity of the electrode tip to a node of Ranvier, occur in only a few per cent of unitary recordings. The ability to record from single nerve fibers is very much dependent on the electrical impedance of the microelectrode: before inserting the electrode through the skin it will have an impedance in the order of MOhms, but as some of the insulating material shears off as the electrode passes through different tissue layers the impedance at the electrode tip will be reduced to the order of kOhms, with the optimal range being some 100 to 500 kOhms. Figure 9.1 shows an example of a recording from a large-diameter myelinated axon—a cutaneous afferent responding to a constant-force indentation of its receptive field on the tip of the index finger. Typically, these receptors generate a

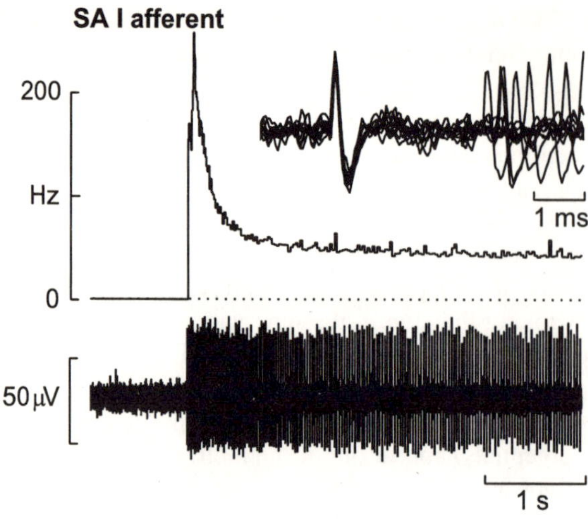

Figure 9.1 Intraneural recording from a slowly adapting cutaneous afferent (type SA 1) via a tungsten microelectrode inserted percutaneously into the median nerve at the wrist. Superimposed action potentials (inset) illustrate the unitary integrity of the recording.

high-frequency discharge at the onset of the stimulus (dynamic phase), which then subsides to a slowly adapting discharge (static phase).

Close examination of spike morphology reveals that 75% of all single-unit recordings from myelinated axons exhibit a secondary peak either on the rising or falling phase of the spike (Vallbo, 1976; Inglis et al., 1997). Whereas the major component of the spike represents the inward currents occurring at the nearby node of Ranvier, the secondary peak reflects action currents at the next node. The significance of this observation is that the majority of unitary spikes recorded from microelectrodes inserted into human nerves conduct past the electrode tip, i.e. the recorded sensory signals progress towards the central nervous system unimpeded by the presence of the microelectrode (Vallbo, 1976; Inglis et al., 1997). Tungsten microelectrodes record preferentially from large-diameter axons, but stable unitary recordings have also been made from non-myelinated axons (C-fibers) in human peripheral and cranial nerves, both from mechanoreceptive and nociceptive afferents (Nordin, 1990; Torebjörk & Ochoa, 1990; Vallbo et al., 1993), and from sympathetic efferents (Macefield et al., 1994; Macefield & Wallin, 1996). By contrast with spikes recorded from myelinated axons, the initial phase of spikes recorded from single C-fibers in human nerves is always negative.

INTRANEURAL MICROSTIMULATION OF SINGLE SENSORY AXONS

In addition to recording from single nerve fibers, it is possible to apply controlled current pulses down the microelectrode to determine the input–output (stimulus-sensation) function of a single afferent. This approach was first applied to single cutaneous afferents in awake human subjects (Torebjörk et al., 1987), but has since been extended to muscle and joint afferents (Macefield et al., 1990). In practice, one uses the microelectrode to locate a single afferent fiber and then stimulates at this intrafascicular site; this allows the experimenter to first identify the afferent and to record its responses to natural stimulation, before determining how the brain utilizes the sensory information evoked by electrical stimulation of this same axon. The validity of the technique relies on there being a stimulus level at which an elementary sensation is projected to a small area within the innervation territory, and that this 'projected field', and its sensory quality remain constant as the stimulus intensity is increased, until a level is reached at which another distinct sensation and projected field are generated. The quantal nature of the sensations evoked by intraneural stimulation at higher intensities can be explained by the spread of current to adjacent axons, which by virtue of the relatively poor somatotopy within nerve fascicles results in sensory percepts that are referred to sites often remote from the receptive field of the afferent that was excited at the lowest stimulus level. The correspondence between an afferent's receptive field and the projected field (as reported by the subject) is remarkably good, supporting the belief that the same sensory axon that can be recorded from the microelectrode is the same axon that is first stimulated electrically.

CONCLUDING REMARKS

It is beyond the scope of this chapter to review all that has been learnt from human microneurography, merely to provide an outline of the approach and a taste of what is possible with this technique. It is recommended that this technique be undertaken with expert guidance. Although there have been no reports of permanent nerve damage in almost thirty years of microneurography in laboratories in Sweden, Australia, Japan and the US, the technique does require skill, patience, and (of course) the approval of your institutional ethics committee.

REVIEW

An excellent review of microneurography has been written by Vallbo et al. (1979), the first generation exponents of the technique. In addition to describing the methodology, it provides examples of its applications to neurophysiology over the first decade. Unfortunately such an all-encompassing review of further progress has not yet been written, although the cited references do indicate some of the scope of its application.

REFERENCES AND FURTHER READING

Edin, B. B. (1990) Finger joint movement sensitivity of non-cutaneous mechanoreceptor afferents in the human radial nerve. *Exp. Brain Res.* **82**, 417–422.

Hagbarth, K. E. and Vallbo, Å. B. (1969) Single unit recordings from muscle nerves in human subjects. *Acta Physiol. Scand.* **76**, 321–334.

Inglis, J. T., Leeper, J. B., Burke, D. and Gandevia, S. C. (1997) Morphology of action potentials recorded from human nerves using microneurography. *Exp. Brain Res.*

Johansson, R. S., Trulsson, M., Olsson, K. A. and Abbs, J. H. (1988) Mechanoreceptive afferent activity in the infraorbital nerve in man during speech and chewing movements. *Exp. Brain Res.* **72**, 209–214.

Knibestol, M. and Vallbo, Å. B. (1970) Single unit analysis of mechanoreceptor activity from the human glabrous skin. *Acta Physiol. Scand.* **80**, 178–195.

Macefield, G., Gandevia, S. C. and Burke, D. (1990) Perceptual responses to microstimulation of single afferents innervating joints, muscles and skin of the human hand. *J. Physiol.* **429**, 113–129.

Macefield, V. G., Gandevia, S. C., Bigland-Ritchie, B., Gorman, R. and Burke, D. (1993) The firing rates of human motoneurones voluntarily activated in the absence of muscle afferent feedback. *J. Physiol.* **474**, 429–443.

Macefield, G., Hagbarth, K. -E., Gorman, R., Gandevia, S. C. and Burke, D. (1991) Decline in spindle support to alpha motoneurones during sustained voluntary contractions. *J. Physiol.* **440**, 497–512.

Macefield, V. G. and Wallin B. G. (1996) Discharge behaviour of single sympathetic neurones innervating human sweat glands. *J. Auton. Nerv. Syst.* **61**, 277–286.

Macefield, V. G., Wallin, B. G. and Vallbo, Å. B. (1994) The discharge behaviour of single vasoconstrictor motoneurones in human muscle nerves. *J. Physiol.* **481**, 799–809.

Nordin, M. (1990) Low-threshold mechanoreceptive and nociceptive units with unmyelinated (C) fibers in the human supraorbital nerve. *J. Physiol.* **426**, 229–240.

Torebjörk, H. E. and Ochoa, J. L. (1990) New method to identify nociceptor units innervating glabrous skin of the human hand. *Exp. Brain Res.* **81**, 509–514.

Torebjörk, H. E., Vallbo, Å. B. and Ochoa, J. L. (1987) Intraneural microstimulation in man. Its relation to specificity of tactile sensations. *Brain* **110**, 1509–1529.

Trulsson, M., Johansson, R. S. and Olsson K. A. (1992) Directional sensitivity of human periodontal mechanoreceptive afferents to forces applied to the teeth. *J. Physiol.* **447**, 373–389.

Vallbo, Å. B. (1976) Prediction of propagation block on the basis of impulse shape in single unit recordings from human nerves. *Acta Physiol. Scand.* **97**, 66–74.

Vallbo, Å. B., Hagbarth, K. -E., Torebjörk, H. E. and Wallin, B. G. (1979) Somatosensory, proprioceptive and sympathetic activity in human peripheral nerves. *Physiol. Rev.* **59**, 919–957.

Vallbo, Å. B., Olausson, H., Wessberg, J. and Norrsell, U. (1993) A system of unmyelinated afferents for innocuous mechanoreception in the human skin. *Brain Res.* **628**, 301–304.

Westling, G. and Johansson, R. S. (1987) Responses in glabrous skin mechanoreceptors during precision grip in humans. *Exp. Brain Res.* **66**, 128–140.

CHAPTER 10

SOMATOSENSORY CORTICAL EVOKED POTENTIALS RECORDED FROM HUMAN SUBJECTS

V. G. Macefield
Prince of Wales Medical Research Institute, Randwick, Sydney

INTRODUCTION

The ongoing electrical activity of the brain can be recorded using surface electrodes glued to the scalp. The resultant electroncephalogram (EEG) is characterized by a mixed frequency, low amplitude signal that, in subjects at rest and with eyes closed, is considered to reflect the summed activity of many cortical neurons firing independently. When the eyes are opened this activity is replaced by a 8–12 Hz signal, the so-called alpha rhythm. This has been assumed to represent the brain's processing of the visual world, but there is strong experimental evidence to support the idea that it reflects nothing more than the electrical activity of the eye muscles (Lippold, 1973). Only during deep sleep (stages 3–4) are the cortical elements of the brain said to be 'synchonized' and then the EEG is characterized by a low-frequency, high-amplitude, signal. However, it is possible to generate synchronized signals in the awake brain of human subjects by using the simple processing technique known as signal averaging. The aim of this approach is to deliver a fixed stimulus to the brain, say by electrically stimulating the median nerve at the wrist once every 3 seconds, and to use a computer to average the electroencephalographic activity recorded from the scalp within a time window extending from some 20 ms before the stimulus to 100 ms or more after the stimulus. After averaging a few hundred sweeps (usually at least 250) a clear negative potential (termed the N20) and positivity (P22) will appear in the averaged records over the cortical receiving area of sensory signals from the hand. These are the earliest components of the 'cortical evoked potential': N20 represents the arrival of the sensory volley at area 3b in the postcentral gyrus, P22 the arrival of the volley at area 4 in the precentral cortex (Maugiere et al., 1983; Desmedt et al., 1987a). It is known from studies in patients with specific lesions of the thalamus that these earliest cortical components are abolished (Maugiere et al., 1983).

RECORDING SOMATOSENSORY EVOKED POTENTIALS

Small Ag-AgCl surface electrodes, filled with conductive gel, can be glued to the scalp (or small stainless-steel needle electrodes can be inserted subdermally) in an arrangement that allows specific cortical areas to be covered, using the international 10–20 system of electrode placement. The cortical receiving area for the hand corresponds to a location approximately 7 cm contralateral and 1 cm posterior to the vertex (the centre of the scalp, measured between nasion and inion and transecting the interaural line). A reference electrode is applied to one or both (electrically linked) ear lobes, which allows small subcortical (far-field) components to be detected in the averaged evoked potential, or a simple bipolar recording arrangement can be used (contralateral scalp electrode referenced to an ipsilateral scalp electrode), which favors the detection of near-field (cortical) electrical activity. Because the signals are small, the amplification gain needs to be at least 1000x, with a band-pass of approximately 0.1–1000 Hz.

EXAMINING CORTICAL PROJECTIONS FROM SKIN AND MUSCLE

The median nerve at the wrist is a mixed nerve—it conveys sensory fibers from a few muscles in the hand as well as most of its skin. It is simple to stimulate cutaneous afferents in the digital nerves of the hand: two saline-soaked pipe cleaners wrapped around the distal phalanx of a finger will do. Selective electrical stimulation of muscle afferents is more tricky, requiring intramuscular microstimulation through a partially insulated tungsten microelectrode inserted at the muscle's motor point. This technique, which allows selective stimulation of muscle afferents without activation of cutaneous afferents, has been used to electrically stimulate low-threshold sensory nerve fibers (i.e. muscle spindle and Golgi tendon organ afferents) from a variety of human skeletal muscles, such as those from the hand, arm and foot (Gandevia & Burke, 1988; Macefield et al., 1989; Gandevia & Burke, 1990), and has also been applied to the human intercostal and parasternal muscles (Gandevia & Macefield, 1989). Comparisons of the scalp distributions of the cortical potentials evoked by stimulation of cutaneous and muscle afferents do reveal differences in the area of cortex to which each species of afferent projects (Gandevia & Macefield, 1989; Macefield et al., 1989; Gandevia & Burke, 1990). One can also use mechanical stimulation over muscle, but this does not activate muscle afferents exclusively. Nevertheless, the fact that stretch reflexes can be generated in the muscle by the same stimulus does mean that muscle spindles are being activated (Macefield & Gandevia, 1992).

APPLICATION OF EVOKED POTENTIALS

Apart from measurement of cortical latencies to peripheral nerve stimulation, which has important clinical application in the diagnosis of certain neurological disorders, somatosensory evoked potentials have been used to study the neural processing of sensory inputs. For instance, comparing the cortical evoked potentials to identical stimuli given in selective attention tasks has revealed the earliest electrical signs of information processing: positive potentials occurring at 40 ms (P40) and 100 ms (P100) and an intervening negativity (N60) at 60 ms (Desmedt et al., 1983, 1987b). Cortical evoked potentials have been used to assess changes in the 'strength' of central synapses (Macefield & Burke, 1991; Gandevia and Ammon, 1992), to study the involvement of the cerebral cortex in long-latency reflex responses to stimulation of muscle or cutaneous afferents (Conrad et al., 1984; Deuschl et al., 1989; Macefield & Johansson, 1994), and to examine task-dependent changes in cortical processing of sensory information during manipulation (Knecht et al., 1993; Macefield & Johansson, 1994).

An example of the latter is illustrated in Figure 10.1 which shows the cortical evoked potential recorded from over the hand area of the scalp during restraint of a manipulandum held between finger and thumb; the subject's task was simply to prevent the manipulandum from slipping out of their grasp. It can be seen that a cortical negativity and subsequent positivity are the first components recorded on the scalp, and these are evoked by excitation of tactile afferents in the finger and thumb by the pulling stimulus. This stimulus generated an automatic increase in the EMG (recorded from the first dorsal interosseous muscle) and consequent increase in grip force; as indicated in the position record, this increase in grip force prevented the manipulandum from escaping. Preceding the subject's automatic response to the pulling stimulus is a large negativity that reflects further cortical processing of the sensory input whereas later potentials represent processing of sensory information resulting from the movement.

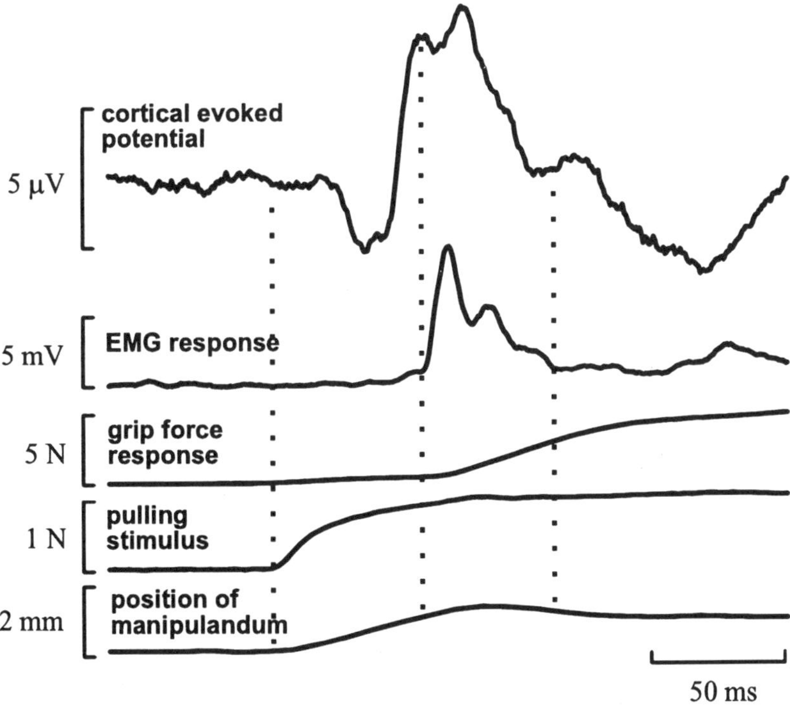

Figure 10.1 Averaged cortical evoked potentials (composed of 517 trials) recorded over the contralateral cortex during restraint of a manipulandum held between finger and thumb. The first vertical line represents the onset of the mechanical stimulus, the second the EMG response to the stimulus. Cortical potentials (negativity upwards) are generated by the stimulus and during the subsequent motor response.

MOVEMENT-RELATED POTENTIALS

Cortical events can also be measured without using sensory stimuli to synchronize the ongoing cortical activity. Instead the subject's own voluntary movements are used to provide the trigger point for synchronizing the averaging. In this case a longer pre-trigger period, of approximately 2 s, and a post-trigger period of some 200 ms, is recorded. Subjects are asked to perform a simple (or sequence of) motor task(s) in their own time, say once every 5–10 s; when the EMG of the relevant muscle crosses a defined electrical threshold the computer starts averaging the cortical activity. Fewer sweeps are required for generating a movement-related potential than for a sensory evoked potential (in the order of 50), and what is observed is a slow increase in negativity over the scalp, termed the 'Bereitschaftspotential' (readiness potential), that develops 1–1.5 seconds before a voluntary, self-paced, muscular contraction (Neshige et al, 1988; Macefield & Gandevia, 1990; Tarkka & Hallett, 1990). This increase in negativity reflects activation of cortical regions in preparation for the manoeuvre, and includes the supplementary motor area and premotor cortical areas. Terminating this slow rise in potential is a pronounced acceleration of this negativity over the primary motor cortex, believed to represent emission of the descending volley from the motor cortex. Following this 'motor potential' is a large positivity and subsequent negativity thought to reflect afferent activity within the cerebral cortex, and the arrival of sensory signals from the contracting muscles and associated tissues (Neshige et al, 1988; Tarkka and Hallett, 1990).

RECENT PAPER

In an interesting paper Kunesch et al (1995) combined microneurography and microstimulation of human nerve fibers with evoked potential techniques. The authors stimulated single cutaneous afferents and recorded the cortical evoked potentials, demonstrating that activation of a single sensory nerve fiber can generate significant activity in cortical neuronal elements.

REFERENCES AND FURTHER READING

Conrad, B., Dressler, D. and Benecke, R. (1984) Changes of somatosensory evoked potentials in man as correlates of transcortical reflex mediation? *Neurosci. Lett.* **46**, 97–102.

Desmedt, J. E., Huy, N. T. and Bourguet, M. (1983) The cognitive P40, N60 and P100 components of somatosensory evoked potentials and the earliest electrical signs of sensory processing in man. *Electroenceph. Clin. Neurophysiol.* **56**, 272–282.

Desmedt, J. E., Nguyen, T. H. and Bourguet, M. (1987a) Bit-mapped color imaging of human evoked potentials with reference to the N20, P22, P27 and N30 somatosensory responses. *Electroenceph. Clin. Neurophysiol.* **68**, 1–19.

Desmedt, J. E., Tomberg, C., Zhu, Y. and Nguyen, T. H. (1987b) Bit-mapped scalp field topographies of early and late cognitive components to somatosensory (finger) target stimuli. *Electroenceph. Clin. Neurophysiol.* (Suppl) **40**, 170–177.

Deuschl, G., Ludolph, A., Schenck, E. and Lucking, C. H. (1989) The relations between long-latency reflexes in hand muscles, somatosensory evoked potentials and transcranial stimulation of motor tracts. *Electroenceph. Clini. Neurophysiol.* **74**, 425–430.

Gandevia, S. C. and Ammon, K. (1992) Selective temporal shift in the somatosensory evoked potential produced by chronic stimulation of the human index finger. *Exp. Brain Res.* **88**, 219–223.

Gandevia, S. C. and Burke, D. (1988) Projection to the cerebral cortex from proximal and distal muscles in the human upper limb. *Brain* **111**, 389–403.

Gandevia, S. C. and Burke, D. (1990) Projection of thenar muscle afferents to frontal and parietal cortex of human subjects. *Electroencephal. Clin. Neurophysiol.* **77**, 353–361.

Gandevia, S. C. and Macefield, G. (1989) Projection of low-threshold afferents from human intercostal muscles to the cerebral cortex. *Respir. Physiol.* **77**, 203–214.

Knecht, S., Kunesch, E., Buchner, H. and Freund, H. J. (1993) Facilitation of somatosensory evoked potentials by exploratory finger movements. *Exp. Brain Res.* **95**, 330–338.

Kunesch, E., Knecht, S., Schnitzler, A., Tyercha, C., Schmitz, F. and Freund, H. J. (1995) Somatosensory evoked potentials elicited by intraneural microstimulation of afferent nerve fibers. *J. Clin. Neurophysiol.* **12**, 476–487.

Macefield, G. and Burke, D. (1991) Long-lasting depression of central synaptic transmission following prolonged high-frequency stimulation of cutaneous afferents: a mechanism for post-vibratory hypaesthesia. *Electroenceph. Clin. Neurophysiol.* **78**, 150–158.

Macefield, G. Burke, D. and Gandevia, S. C. (1989) The cortical distribution of muscle and cutaneous afferent projections from the human foot. *Electroenceph. Clin. Neurophysiol.* **72**, 518–528.

Macefield, G. and Gandevia, S. C. (1990) The cortical drive to human respiratory muscles in the awake state assessed by premotor cerebral potentials. *J. Physiol.* **439**, 545–558.

Macefield, G. and Gandevia, S. C. (1992) Peripheral and central delays in the cortical projections from human truncal muscles. Rapid central transmission of proprioceptive input from the hand but not the trunk. *Brain* **115**, 123–135.

Macefield, V. G. and Johansson, R. S. (1994) Electrical signs of cortical involvement in the automatic control of grip force. *Neuroreport* **5**, 2229–2232.

Mauguiere, F., Desmedt, J. E. and Courjon, J. (1983) Neural generators of N18 and P14 far-field somatosensory evoked potentials studied in patients with lesion of thalamus or thalamo-cortical radiations. *Electroencephal. Clin. Neurophysiol.* **56**, 283–292.

Neshige, H. Lüders, L. and Shibasaki, H. (1988) Recording of movement-related potentials from scalp and cortex in man. *Brain* **111**, 719–736.

Tarkka, I. M. and Hallett, M. (1990) Cortical topography of premotor and motor potentials perceding self-paced, voluntary movement of dominant and non-dominant hands. *Electroenceph. Clin. Neurophysio.* **75**, 36–43.

SECTION 4
APPLICATION OF DRUGS

CHAPTER 11

VERY FAST DRUG APPLICATION TO DISSOCIATED CELLS AND MEMBRANE PATCHES

J. Clements
Division of Neuroscience, John Curtin School of Medical Research, Australian National University

INTRODUCTION

Synaptic transmission can only be fully understood when we know about the kinetic properties of transmitter release and clearance, and about the kinetic interactions between transmitters and receptors. The voltage-clamp technique has been used extensively to study the kinetics of voltage gated channel activation and inactivation (e.g. sodium and potassium channels), but until recently there was no equivalent technique for studying the kinetics of transmitter gated channels (receptors). This situation was remedied with the development of the fast solution exchange technique (sometimes referred to as concentration-clamp). Fast solution exchange has revolutionized the study of synaptic transmission, and promises to remain an important neuroscience research tool. This chapter will outline the three main experimental methods for achieving very rapid solution exchange at the surface of dissociated cells and outside-out membrane patches.

THREE FAST SOLUTION EXCHANGE METHODS

Piezo, Outside-out Patch

A two-barrel flow pipe mounted on a piezoelectric translator allows very fast (<1 ms) solution exchange at the surface of an outside-out membrane patch (Figure 11.1). Control and drug solutions both flow continuously from adjacent barrels of the flow pipe. An outside-out membrane patch is pulled from a cell, then the patch electrode is lifted away from the tissue and its tip positioned in the control solution flow close to the interface with the test solution. The translator moves the flow pipe so that the solution interface sweeps past the stationary membrane patch. Only a small translation is required (~20 μm), and this can be completed very quickly.

Piezo, Whole-Cell

A similar arrangement can achieve fast (<10 ms) solution exchange at the surface of an acutely dissociated or cultured cell (Figure 11.2). Because a cell is larger than a

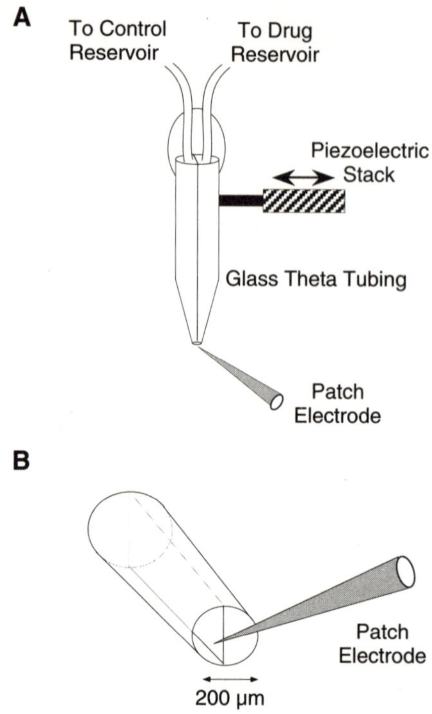

Figure 11.1 A: A flow pipe constructed from glass theta tubing is shown schematically. Two fine plastic tubes are glued into the two barrels of the theta tubing at the top of the flow pipe. These plastic tubes continuously feed two different solutions into the flow pipe. The bottom of the flow pipe is tapered to a small opening (200 µm diameter). The flow pipe is mounted on a piezo-electric translator which moves it from right to left. B: A patch electrode is positioned near the tip of the flow pipe, in the solution flowing from the left barrel. When the flow pipe is moved to the left, the tip of the electrode experiences a rapid solution exchange (<1 ms) as it crosses the interface between the two flowing solutions. This system can be used with outside-out membrane patches, or with small cultured cells that are not firmly attached to a culture dish.

membrane patch and is typically adhering to the base of a culture dish, larger flow pipes are required. Typically, a multi-barrel flow pipe is used, and control and drug solutions flow continuously from adjacent barrels. A large translation is needed to exchange drug solutions (~400 μm).

Solenoid Valve, Whole-Cell

An alternative approach can be used with cultured cells (Figure 11.3). Solenoid valves are used to switch solutions instead of a piezoelectric translator. The multi-barrel flow pipe remains stationary, and flow through one barrel is turned off at the same moment flow in an adjacent barrel is turned on. Because the flow fans out from an individual barrel, there is a region between the two barrels where rapid (<20 ms) solution exchange occurs (Figure 11.3 B).

THEORY OF OPERATION

Piezoelectric Translators

Piezoelectric materials change shape in response to an applied voltage. The maximum dimensional change of a single transducer is on the order of a few micrometers. Multiplication or amplification of this change is required to construct a useful translator. Two common approaches are stacks and bimorphs. Stacks are several piezoelectric transducers connected mechanically in series and electrically in parallel. The total displacement of the stack is equal to the sum of the individual displacements. Bimorphs consist of two piezoelectric plates bonded together so that an applied voltage causes the plates to deform in opposite directions. This opposition causes the bimorph to bend. Bimorphs provide a much larger translation than a stack. They also require lower operating voltages and currents. However, they generate much less force per volt, and are not as stiff or robust as stacks. The speed of translation is limited by the mass of the flow pipe mounted on the translator and by the maximum force the translator can generate. The intrinsic stiffness of the translator and the mass of the flow pipe determine the amplitude and persistence of resonant ringing following a rapid translation. For these reasons, piezoelectric stacks provide faster translation speed and fewer problems with ringing than bimorphs.

Solenoid Valves

Solenoid valves are activated by a pulse of current through a coil. This generates a magnetic field that moves an internal plunger, thereby closing or opening the valve. They require a pulse of high current (up to 1 amp) for rapid valve operation. A custom power supply (or valve controller) provides this current. Solenoid valves come in two standard configurations: a two-way valve with only a single outlet, or a three-way valve that switches flow between two outlets. Although the option of switching between outlets is rarely required, three-way valves are preferable because they produce a smaller pressure pulse when switched. In contrast, the plunger movement in a two-way valve violently expels a small volume of solution as it closes. This can literally blow away a cell positioned in front of the flow pipe.

FABRICATION OF FLOW PIPES

Flow pipes can be constructed from specialized multi-barrel glass tubing, or by gluing together an array of individual tubes. The simplest multi-barrel glass tubing is circular with a central septum, usually called theta tubing (Figure 11.1B). Custom multi-barrel tubing can be constructed by sintering an array of square glass tubing. When working

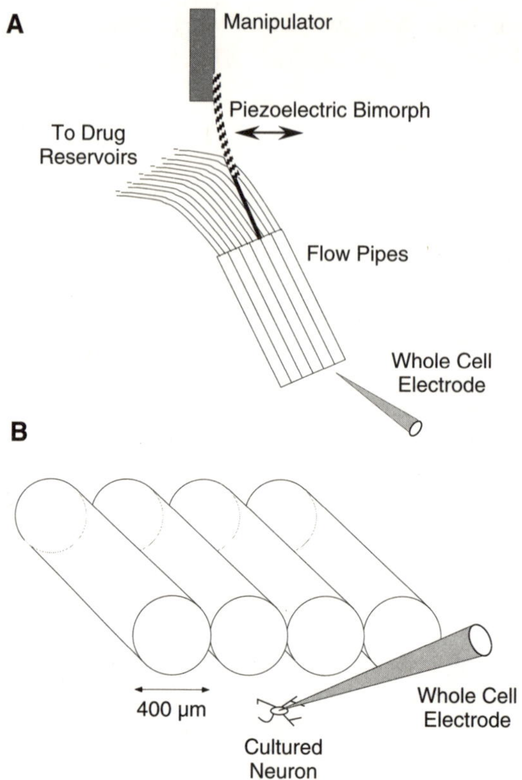

Figure 11.2 A: Several quartz flow pipes are glued together to form an array. A plastic tube is connected at the top of each pipe to supply a flow of drug solution. The array is mounted on a piezoelectric bimorph which moves it laterally. B: Each tube is approximately 400 μm in diameter, sufficient to perfuse the dendritic tree of a cultured neuron. It is positioned so that one pipe points directly at a cultured cell. When the bimorph moves the array so that a different flow pipe points towards the cell, a moderately fast solution exchange is achieved (10–100 ms).

with membrane patches, a multi-barrel glass flow pipe is typically pulled down to an outlet diameter of 100–400 μm total (50–200 μm per barrel). The multi-barrel glass is pulled in a conventional microelectrode puller. A high temperature setting is used to give a long tapered tip. The tip is then scratched at the required diameter using a sharp diamond pencil under a dissection microscope (steady hands required). The tip is bent gently until it breaks at the scratch line. When choosing the flow pipe tip diameter there is a trade off is between the thickness of the lumen (the thinner the lumen the faster the solution exchange), and the tendency of thinner barrels to become blocked by small particles. Ordinary polyethylene tubing can be used to connect from solution reservoirs to the back of the flow pipe. Flow pipe life time is limited due to leaks and blockages.

A hand fabricated array of individual tubes is generally used for whole cell perfusion. Circular glass or quartz tubing is available in a range of diameters down to about

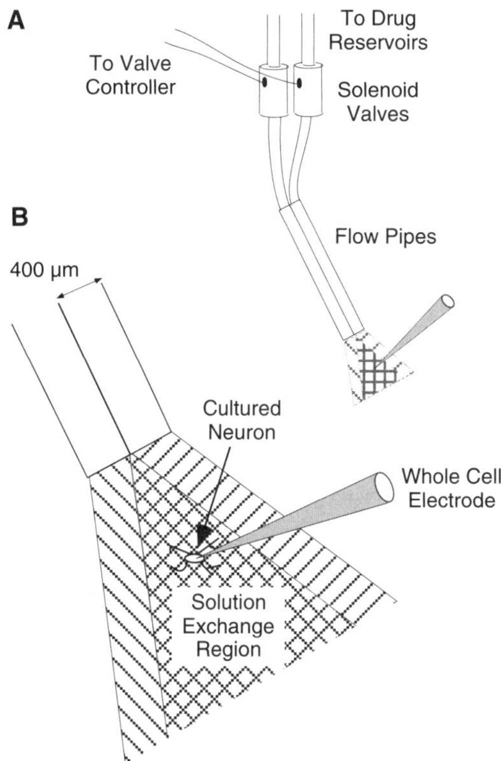

Figure 11.3 A: Two glass or quartz flow pipes are glued together and a plastic tube is connected at the top of each pipe to supply drug solution. The tubes are connected via solenoid valves which can turn the solution flow on and off. Solution only flows into one pipe at a time. B: The pipes are positioned above a cultured cell so that the solution flowing from one flow pipe fans out over the cell in a broad V shape. When the solenoid valves are activated, solution from the second pipe again fans out over the cell, displacing the first solution. This system delivers moderately fast solution exchange (~10 ms).

a hundred micrometers, and can be glued together in any configuration. Plastic coated quartz tubing is much more robust than glass tubing. A horizontal array of 5 to 10 parallel tubes each of about 400 μm is common (B in Figures 11.2 and 11.3).

USING A FAST SOLUTION EXCHANGE SYSTEM

Problems with Resonant Ringing of Piezoelectric Translators

A potential problem, especially with a bimorph translator, is residual vibration (resonant ringing) following a rapid translation. This can be effectively eliminated by ramping the command voltage rather than stepping it. A simple way to do this is to add a series resistance to the control circuit. The piezoelectric translator has an intrinsic capacitance which interacts with the series resistance to determine the time constant of the voltage rise at the translator. For a bimorph, a resistor of 100–500 kOhm should slow the rise sufficiently to eliminate ringing. A more sophisticated approach is to use a power supply capable of generating a true voltage ramp. Unfortunately, ramping the voltage will reduce the maximum translation velocity and slow the solution exchange. This places a fundamental limitation on the briefest pulse of test solution that can be applied to a membrane patch using a bimorph. In practice, pulse lengths as brief as 10 ms can still be obtained.

Piezoelectric stacks are much stiffer and faster than bimorphs, and less prone to ringing after a rapid translation. Even so there are still some problems with high frequency ringing due to the flexibility and elasticity of the flow pipe. Shortening the flow pipe or using thicker glass tubing may reduce this problem. Using a stack translator it is possible, under ideal conditions, to apply a pulse of test solution as brief as 0.5 ms.

Measuring Fast Solution Exchange

When setting up and fine tuning a fast solution exchange system it is vital to monitor the exchange time course at the open tip of an electrode. This is the best way to practice correct placement of the electrode relative to the flow pipe. It can also be used to test for residual vibration of a translator. To monitor solution exchange, fill the test barrel of the flow pipe with bath solution diluted by 20–50% with distilled water. This produces a big tip potential offset at the open tip of the recording electrode. The change in tip potential accurately reflects the solution exchange time course. Use an ordinary patch recording electrode (tip diameter 1–3 μm), and apply some positive pressure. Without the positive pressure the tip potential responds slowly following solution exchange, presumably because of diffusion of the diluted test solution into the tip of the recording electrode.

The Importance of Shielding

Piezoelectric translators work at high voltages (10 to 1 000 V), and tend to produce big electrical artifacts in electrophysiological recordings. To reduce these artifacts, surround the translator and the wires to the voltage supply with an electrically conducting shield (e.g. aluminium foil) and connect this to the table earth. Stack translators are sometimes provided with shielded cabling, but it may still be helpful to earth the cable's shield.

A Few Hints

Filter all solutions and use dust covers over solution reservoirs to reduce the chance of flow pipe blockage. Warm all solutions to above room temperature before the start of an experiment to de-gas them, and reduce the chance of bubbles blocking the flow pipe. After each fast solution exchange recording, it is a good idea to perform the following control. Before starting the experiment, dilute the test solution by 2% relative to control with distilled water—just enough to produce a detectable change in open tip potential, but not enough to distort the recording when a membrane patch or cell is present. At the end of the recording, blow away the patch or cell with a pressure pulse, leave a small positive pressure on the pipette, and record some open tip responses with the pipette in the same location as used during the recordings. This defines the exact time of solution exchange, and ensures that the flow pipe was correctly positioned and there was no ringing.

REFERENCES AND FURTHER READING

Corey, D. P. and Hudspeth, A. J. (1980) Mechanical stimulation and micromanipulation with piezoelectric bimorph elements. *J. Neurosci. Meth.* **3**, 183–202.

Franke, C., Hatt, H. and Dudel, J. (1987) Liquid filament switch for ultra-fast exchanges of solutions at excised patches of synaptic membrane of crayfish muscle. *Neurosci. Lett.* **77**, 199–204.

Lester, R. A. J., Clements, J. D., Westbrook, G. L and Jahr, C. E. (1990) Channel kinetics determine the time course of NMDA receptor-mediated synaptic currents. *Nature* **346**, 565–567.

Colquhoun, D., Jonas, P. and Sakmann, B. (1992) Action of brief pulses of glutamate on AMPA/kainate receptors in patches from different neurons of rat hippocampal slices. *J Physiol.* **458**, 261–87.

CHAPTER 12

MICROELECTROPHORESIS AND PRESSURE EJECTION METHODS

G. Lacey
Therapeutic Goods Administration, Commonwealth Department of Health and Family Services, Canberra

INTRODUCTION

Microelectrophoresis (sometimes called iontophoresis or ionophoresis) was developed in the 1950s as a method for applying pharmacologically active compounds to very limited areas of muscle or nerve cell membranes. The technique involves the ejection of active compounds from ionized solutions contained within micropipettes by the use of electrical current. The technique of microelectrophoresis can be easily combined with recording of intracellular or extracellular electrical activity from the cells under investigation.

BASIC PRINCIPLES

Micropipettes with small orifices, usually less than 2.0 μm, are filled with aqueous solutions of the pharmacologically active compound. Electrical contact with the drug solution is made by inserting a fine silver wire into the micropipette. By establishing a potential difference between the drug solution and the pipette surroundings, ions will move through the solution and out of the pipette tip.

If the pipette is made positive to the surrounding medium, an outward current will flow and positively charged ions (cations) will be ejected from the pipette (Figure 12.1A). This outward current is sometimes referred to as a cationic current as it ejects cations. By reversing the direction of the current, negatively charged ions (anions) will be ejected (Figure 12.1 B).

As drug molecules would tend to diffuse from the micropipette tip, it is necessary to apply a small current of appropriate direction, known as the retaining current, to counteract this diffusion. This retaining current is sometimes referred to as the backing, braking or holding current.

The use of a retaining current to prevent the drug diffusing from the micropipette will obviously cause ejection of one of the ions in the solution. It is therefore important to determine that the ion of opposite charge to the drug is pharmacologically inert.

The voltage required to pass current through a micropipette in order to eject and retain a compound is determined by the electrical resistance of the pipette. The electrical resistance of a micropipette depends upon many things including the physical dimensions of the micropipette, most importantly the orifice size, the concentration of the drug in solution and its degree of ionisation. The electrical resistance of micropipettes can be easily

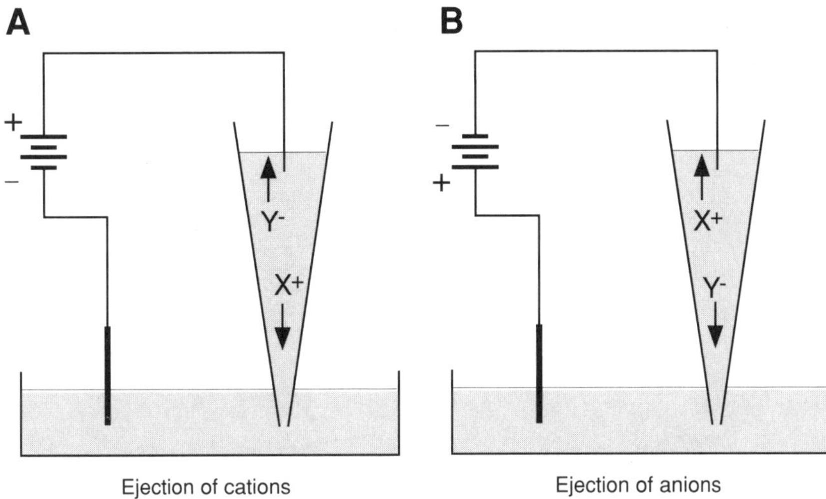

Figure 12.1 Ions are ejected or retained within the micropipette depending upon the potential difference across the pipette tip.

measured and therefore the voltage required to pass the desired current can be calculated by Ohm's Law. For example, to pass 100 nA of current through a 10 MOhm pipette will require 1V. By monitoring both the voltage required and the current passed through the microelectrophoretic pipette throughout the experiment, the investigator is aware of any changes in electrical resistance of the pipette. Such changes in resistance will indicate that the microelectrophoretic pipette has become blocked or broken and should be replaced.

The amount of retaining current needed to prevent diffusion will also depend on the micropipette resistance, but as a rule a potential difference of 500 mV should be maintained between the tip of the pipette and its surroundings.

TYPES OF MICROPIPETTE

Depending on the experimental system being used, different types of microelectrophoretic pipette are used. For example when drugs are ejected onto the neuromuscular junction in an isolated muscle preparation or onto single cells in culture, single barrel pipettes can be used. These single barrel pipettes can be positioned under visual control by a separate micromanipulator close to the recording electrode. However in situations where recordings are made deep into tissue of the central nervous system, the microelectrophoretic pipette must be joined to the recording electrode so that they can be positioned with the same micromanipulator. Various types of pipette have been developed for each purpose, the most useful of these being the multibarrel pipettes whereby an extracellular recording electrode is surrounded by several drug containing pipettes and

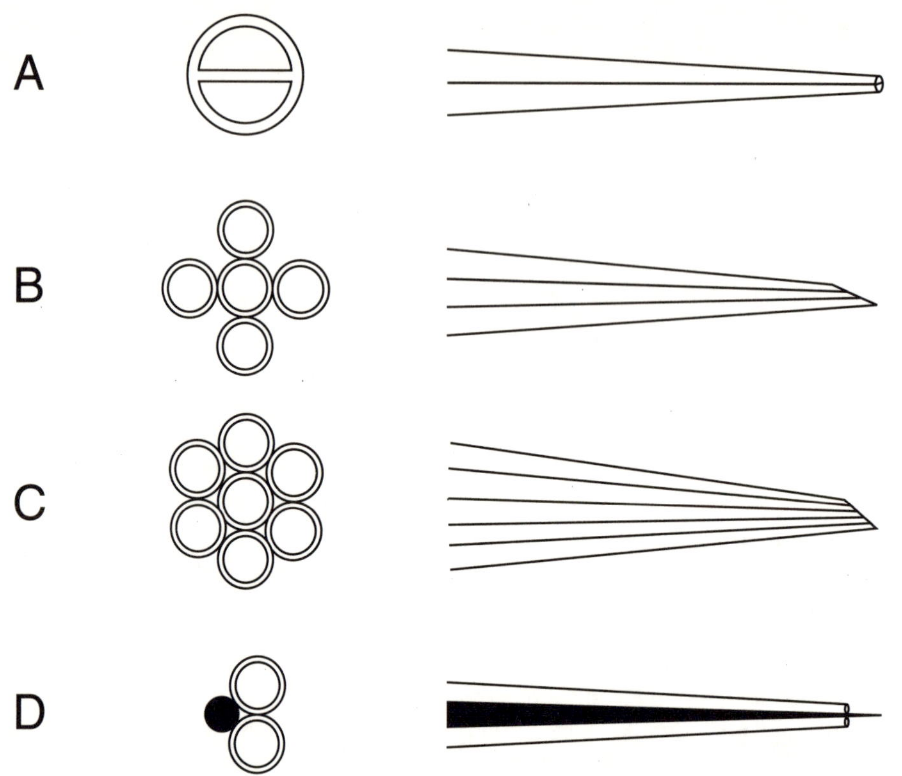

Figure 12.2 Examples of pipette assemblies for microelectrophoresis. A: Theta glass; B: Five-barrel assembly; C: Seven-barrel assembly; D: Double-barrel assembly with attached intracellular electrode.

fused together to make a multibarrel assembly with a total tip diameter of less than 10 μm. It is also possible to attach an intracellular microelectrode to such an assembly.

Most pipettes are made from borosilicate glass capillaries fused together. The glass is then melted and pulled to a fine tip by an electrode puller. Multibarrel assemblies are often ground to a bevelled tip to aid penetration of tissue.

CURRENT BALANCING

In situations where microelectrophoretic pipettes are placed close to excitable membranes, the passage of current itself maybe sufficient to influence membrane excitability. To overcome this problem a technique called current balancing can be used. One barrel of a multibarrel assembly is filled with an inert solution such as sodium chloride, and throughout the experiment current equal and opposite to the algebraic sum of the currents in all the other barrels is constantly passed through this sodium chloride containing barrel. In this way the net current flow at the tip of the pipette is kept as close as possible to zero.

FILLING SOLUTIONS

It is essential that all filling solutions be centrifuged or filtered before use as the presence of any small particles in a microelectrophoretic pipette will rapidly lead to blocking of the orifice. It is also important that solutions be made in glass distilled water, as contaminating ions could carry a significant proportion of the electrophoretic current during drug ejection.

If the solution in the pipette has a substantially different osmotic concentration to the surrounding medium, bulk flow of the solvent could result. To prevent this it is desirable that the filling solutions contain ionized compounds at a final concentration of 150–200 mM, which makes them approximately iso-osmolar with extracellular fluid. If the compound of interest is not sufficiently soluble to attain such a concentration, it may be dissolved in 150 mM sodium chloride to achieve the appropriate osmolarity. Also, if the compound of interest has a very high potency it maybe desirable to dissolve it in NaCl so that only a proportion of the electrophoretic current is being carried by the drug, the rest by sodium or chloride ions. In this way reasonable ejecting currents can be used without saturating receptors with drug.

Although many problems can be solved by dissolving drugs in NaCl, one important problem should be borne in mind. If the ion of interest is negatively charged an outward (anionic) retaining current would be needed, and this would cause the ejection of sodium ions. When sodium ions are ejected, they take with them their water of hydration and thus some of the dissolved drug.

Another problem that can be encountered with very potent drugs is that of uncontrollable leakage. In some cases it seems that it is impossible to retain the drug sufficiently with normal retaining currents and if excessive retaining currents are used, subsequent ejection is impaired as the drug has become very dilute at the tip of the pipette. In these cases care must be taken to use the drug in very dilute solutions particularly if the electrophoretic pipette is to be left near the recording site for any length of time.

PROBLEMS ASSOCIATED WITH MICROELECTROPHORESIS

The major problem with microelectrophoresis is that the concentration of a drug achieved within tissue is impossible to quantify accurately. The drug concentration at the membrane of a cell of interest will depend on many factors, including the concentration of the drug within the pipette, the electrophoretic current passed and the proportion of this current which carries the drug. Other factors that will influence the amount of drug reaching the cell are the distance of the pipette from the cell and processes in the extracellular environment which may inactivate the drug. Furthermore as the drug is applied from a point source, its concentration around a cell will not be uniform. It is therefore important that sufficient controls are performed for each experiment and that great care is taken when comparing results between experiments. Simply because 100 nA of current was passed for 1 minute through a 50 MOhm pipette containing a 150 mM solution of drug placed 100 μm from a cell, does not mean that the same parameters used in another experiment will result in the same concentration of drug reaching both cells.

EQUIPMENT

It is not appropriate in this short review to give detailed descriptions of the types of electronic circuitry that can be employed for microelectrophoresis, or of the construction and filling of micropipettes. For those interested, these topics are fully covered by Stone (1985). Equipment is available commercially for controlling and timing of ejection and retaining currents.

PRESSURE EJECTION

It is obvious from the above discussion that only drugs that are ionized in solution can be applied by microelectrophoresis. Pressure can be used to eject solutions of unionized compounds from micropipettes. This pressure can be applied either by a syringe or from a source of compressed gas. Pressure ejection has several disadvantages over microelectrophoresis. Firstly there is no control of diffusion of the drug from the pipette. Secondly, the pipette orifice may become blocked by material within the solution or by external tissue fragments and therefore an increase in pressure will not lead to ejection of the drug. Some investigators try to prevent this problem by using very fine bore capillaries for ejection pipettes and watching under a microscope for changes in level of the filling solution during pressure application. Thirdly, pressure ejection of fluid results in physical displacement of tissue. This displacement can complicate the interpretation of changes in the activity of cells, within the vicinity of the pipette orifice, solely in terms of the drug applied.

Pressure ejection has been used successfully in experimental systems such as recording from cells in culture, where the ejection pipette can be placed in the perfusing solution close to the cell of interest. However if the ejection pipette has to be lowered through a significant amount of tissue as in a slice or *in vivo* preparation, the pipette is much more likely to become blocked.

REFERENCES AND FURTHER READING

Stone, T. W. (1985) *Microiontophoresis and Pressure Ejection*. IBRO Handbook Series: Methods in the Neurosciences, Vol. 8. Chichester: John Wiley & Sons.

SECTION 5

MEASUREMENT OF ION CONCENTRATIONS

CHAPTER 13

ION-SELECTIVE MICROELECTRODES

A. I. Cowan
Division of Neuroscience, John Curtin School of Medical Research, Australian National University

INTRODUCTION

Ion-selective microelectrodes are used to measure the concentration of a specific ion. In neuroscience they are commonly used to measure extracellular ion concentrations and less commonly used to measure intracellular ion concentrations. A number of different ions have been measured using such electrodes. These include sodium, calcium, potassium, hydrogen and chloride ions.

THEORY

Most ion-selective microelectrodes in current use are based on a 'liquid membrane' placed in the tip of a glass microelectrode. This liquid is permeable to a specific ion, but excludes other ions (it is a selective ionophore). The potential recorded by the electrode measures differences in the concentration of that ion between the two sides of the liquid membrane (i.e. the electrode and the external solution). The electrode will also measure other changes in electrical potential, for example when an intracellular recording is made the electrode will record a potential that is a combination of the membrane potential and the potential due to the ion concentration. Thus to measure only a potential associated with the ion concentration one must use a reference electrode that measures all other electrical potentials. When recording extracellularly or when recording from large neurons a separate reference electrode, placed near the ion-selective electrode, can be used. However, when performing intracellular recording from small neurons (e.g. mammalian CNS neurons) only one fine electrode can be used. In this case a double-barrelled electrode is manufactured where one barrel is selective for a particular ion (contains the liquid membrane) and the other is used as the reference barrel (Figure 13.1).

The output of the ion-selective electrode is a voltage, but the information that the experimenter requires is the concentration of the ion. To obtain this information one must perform a calibration of the electrode.

CALIBRATION

Calibration of a microelectrode is performed using a number of solutions with different concentrations of the ion under question. The solutions must contain other ions at the

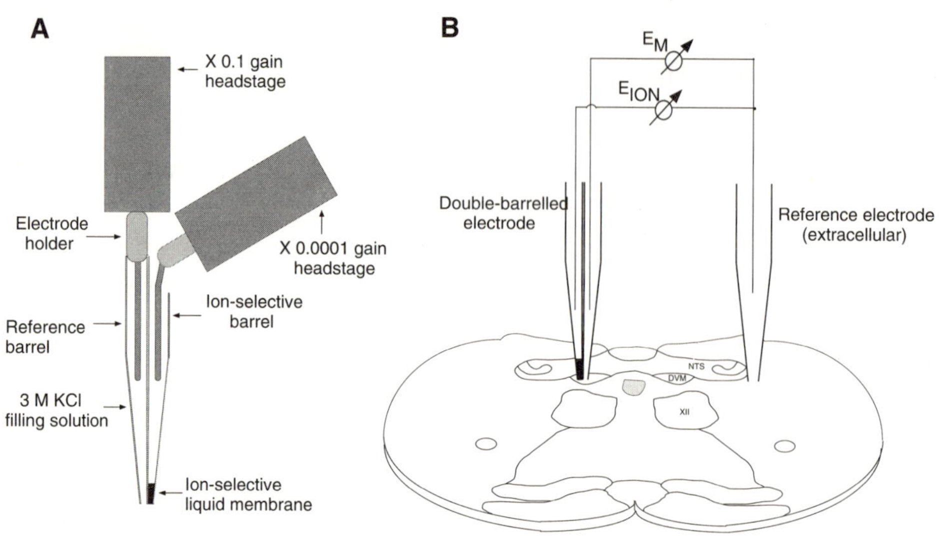

Figure 13.1 A: Schematic diagram of a double-barrelled electrode. B: Diagram of the recording set-up for intracellular recording in a brainstem slice.

concentrations expected to be contained in the experiment. This is because the total ionic concentration and the presence of other ions may affect the potential recorded by the electrode (this is more important for some ions than others, as will be discussed in more detail below). A sample calibration of a pH-selective electrode is shown in Figure 13.2.

In fact the ion-selective electrode measures changes in ion activity rather than concentration, but as long as the total ion activity remains relatively constant and the electrode is specific for only one ion then the electrode can be calibrated in relation to ion concentration rather than activity.

TECHNICAL AND THEORETICAL CONSIDERATIONS

Sensitivity to ion concentration

Theoretically a perfectly selective electrode will behave in a Nernstian fashion, i.e. for a monovalent ion a 10-fold change in concentration there will be a 62 mV change in electrode potential[1]. However such a relationship is not often achieved and the sensitivity of the electrode is much less than this. Armstrong and Garcia (1980) have suggested that: (1) the presence of a low resistance shunt pathway between the glass wall of the electrode and the ionophore solution due to incomplete silanisation[2] may attenuate the potential difference between the external and internal solutions of the microelectrode, and (2) the silane residues attached to the glass can act as cation exchangers and thereby

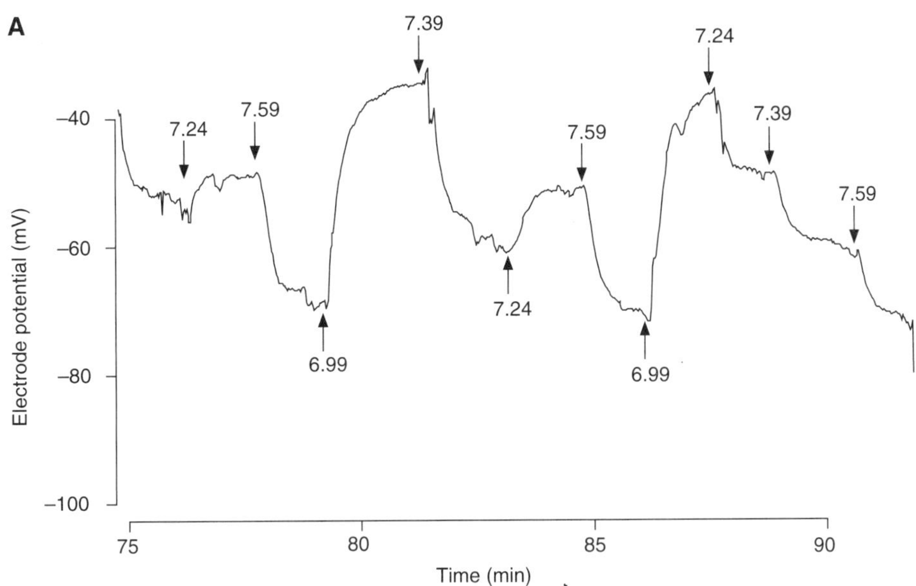

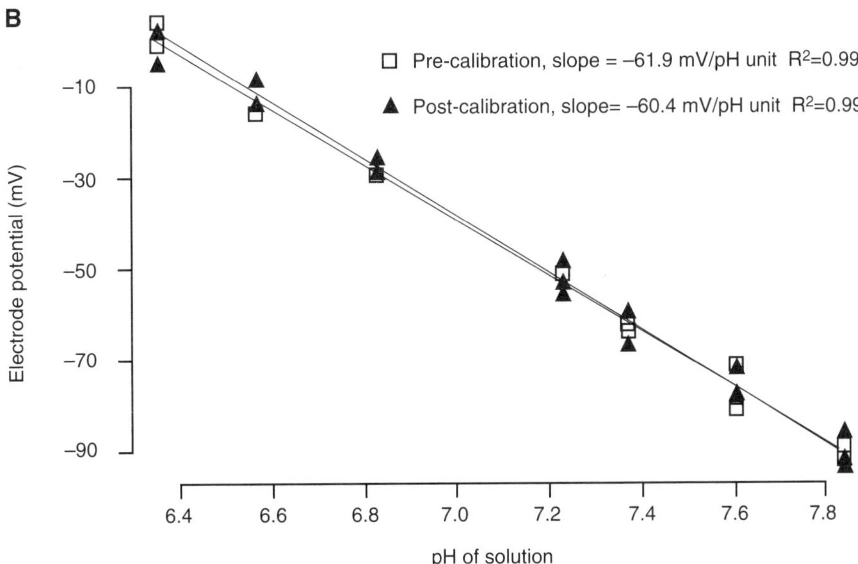

Figure 13.2 A: Voltage trace recorded from a double-barrelled pH-selective microelectrode during a series of solution changes. The numbers on the graph indicate the pH of each new calibration solution. B: Linear regression plots of the calibration of a different pH-selective microelectrode, where the calibrations were done before and after an experimental recording.

modify the sensitivity of the electrode. These factors become more important with decreasing tip-size.

The actual sensitivity of the electrode is not particularly important as long as: (1) the sensitivity of the electrode is much greater than the recording noise and, (2) recordings are only accepted if the sensitivity of the electrode remains unchanged throughout the recording period.

Ion selectivity

If the ionophore is not sufficiently selective for a particular ion then the Nernst equation is not appropriate. In this case calibrations must include those in which the concentration of the interfering ions has been changed and the equation that is then used is the Nikolsky equation[3]. An increasing number of highly selective ionophores are being developed.

Temporal response

Because of the high resistance of the ion-selective barrel (the liquid membrane is a highly viscous hydrophobic material which introduces a large resistance) and the resultant large RC time constant, an ion-selective electrode has a slow response time. This response time may be up to a few seconds. For this reason one must be cautious in interpreting data on a fast time base. For example attempting to measure ion concentrations during synaptic events or during an action potential may not be appropriate.

Cell contamination

Intracellular recording techniques have the potential to disturb normal neuronal function if the contents of the electrode leak into the cell. Leakage of the ionophore into the cytoplasm is unlikely to occur because the ionophore is hydrophobic. However, diffusion of the ionophore via the silanized glass of the electrode into the surrounding plasma membrane could occur and this would create a new ion conducting pathway. Oesch, Ammann and Simon (1987) modelled such a diffusion process and suggested that leakage in this manner would be most significant in those cells that have a relatively high input resistance (as a new ion conducting pathway would result in a proportionately larger increase in conductance than in those cells with a low input resistance).

Electrical coupling

A major problem with double-barrelled electrodes is the electrical coupling between the two barrels (Figure 13.3). Significant electrical coupling would result in an inability to accurately determine what proportion of the electrode response was due to changes in ion concentration and what proportion was due to changes in membrane potential, because any change in potential in one barrel would affect the potential recorded in the other barrel. 'Coupling resistance' is a measure of the coupling between the two barrels[4]. To prevent significant coupling it should be below 250 kOhm. Several techniques are used to reduce electrical coupling.

Thick septum theta tubing

Thick septum theta tubing has several advantages over tubing where the septum is as thin as, or thinner than, the outer wall. These advantages include a decreased likelihood of the septum developing cracks and a potential for extension of the septum beyond the outer glass wall at the tip of the electrode. As a consequence there is improved isolation of the ion pools 'seen' by the two electrode tips (Brown & Flaming, 1986).

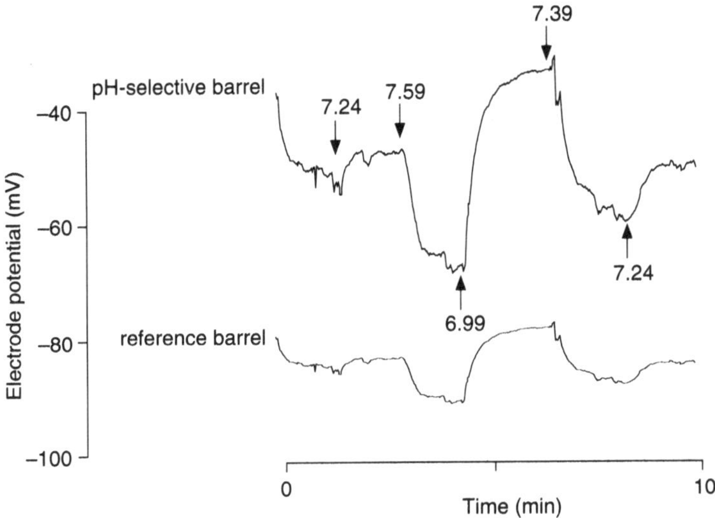

Figure 13.3 In this (simulated) example of a calibration of a pH-selective electrode one can see that the reference barrel of the electrode also shows changes of potential with each change of solution. This is because the two barrels of the electrode are electrically coupled, and the reference barrel 'sees' some of the change in potential in the pH-selective barrel. Such an electrode should not be used in an experiment.

Electrode resistance

The resistive coupling between the two barrels is minimized by keeping the electrical resistance of each barrel low. However this requirement conflicts with the necessity to have fine tipped microelectrodes. A compromise can be reached by using electrodes with a short shank length, and a relatively steep taper.

RECENT PAPER

A good example of a paper where ion-selective electrodes have been used is by Schweitzer and Williamson (1995). In this paper the authors investigate the cause of prolonged nonsynaptic, seizure-like bursts of activity in the dentate granule cell layer of the hippocampus. They show that following repetitive stimulation of afferents the extracellular calcium concentration decreased whereas the extracellular potassium concentration significantly increased. Associated with these changes in ion concentrations was an increase in the electrical activity (bursts of population spikes). Similarly when they perfused the slice with a solution containing potassium and calcium concentrations equivalent to those measured following the repetitive stimulation, they observed a similar increase in electrical activity. From these studies they concluded the changes in extracellular ion concentrations initiated the increase in electrical activity, and that this sort of change may be important in epilepsy.

POINTS TO CONSIDER WHEN INTERPRETING DATA USING ION-SELECTIVE ELECTRODES

1. Is there any evidence of coupling between the two barrels (with a double-barrelled electrode)?
2. How good is the quality of recording? Are the membrane potential, action potential amplitude and input resistance of the cell similar to those recorded with fine single-barrelled electrodes or with the whole-cell patch technique? Consider also if there is evidence of contamination by the ionophore, e.g. a progressive decrease in input resistance.
3. Are the type of events being measured appropriate for the slow time-course of these electrodes?
4. Is the ionophore sufficiently selective for the ion under investigation? If not, were appropriate calibrations performed? Is there some discussion of the possibility that a change in concentration of an interfering ion contributes to the measurement?
5. Is the sensitivity of the electrode appropriate for the size of the measurements being made?
6. Is there any safeguard that the sensitivity of the electrode does not change during the course of the experiment (e.g. calibration both before and after the experiment)?

NOTES

1. The theoretical sensitivity of an ion-selective microelectrode is approximated by the Nernstian Relationship (Moody & Thomas, 1971):

$$\Delta E_{elec} = \frac{RT}{zF} \ln \frac{[\text{ion}]_I}{[\text{ion}]_{II}}$$

where $[\text{ion}]_I$ and $[\text{ion}]_{II}$ are the concentration of the ion in two solutions, R is the Gas constant, F is Faraday's constant, z is the charge on the ion and T is the temperature in Kelvin.
2. Silanisation is a process that makes the walls of the glass hydrophobic. This is important because it allows the hydrophobic ionophore solution to get to the tip of the electrode.
3. A more complete description of the voltage generated by an ion-selective electrode when exposed to a variety of ions is given by the Nikolsky equation:

$$\Delta E_{elec} = E_o + \frac{RT}{z_I F} \ln (a_I + \Sigma K_{I/II} a_{II}^{(Z_I/Z_{II})})$$

where a_I is the activity of the primary ion, and a_{II} is the activity of the interfering ion. Z_I, Z_{II} are the valencies of the primary and interfering ion; $\Sigma K_{I/II}$ is a selectivity coefficient describing the relative degree of interaction of the interfering ion compared to the primary ion with the ion-selective membrane.

4. Coupling resistance (R_c) is determined according to the equation:

$$R_c = \frac{E_2}{I_1}$$

where I_1 is the current passed through one channel and E_2 is the voltage measured across the other channel (Brown & Flaming, 1986).

REFERENCES AND FURTHER READING

Armstrong, W. M. and Garcia, D. J. (1980) Ion-selective microelectrodes: theory and technique. *Fed. Proc.* **39**, 2851-2859.

Brown, K. T. and Flaming, D. G. (1986) *Advanced micropipette techniques for cell physiology*. Chichester: John Wiley and Sons.

Fry, C. H., Hall, S. K., Blatter, L. A. and McGuigan, J. A. (1990) Analysis and presentation of intracellular measurements obtained with ion-selective microelectrodes. *Exp. Physiol.* **75**, 187–198.

Meyerhoff, M. E. and Opdycke, W. N. (1986) Ion-selective electrodes. *Adv. Clinical Chem.* **25**, 1–47.

Moody, G. J. and Thomas, J. D. R. (1971) *Selective ion sensitive electrodes*. Watford: Merrow.

Muckenhoff, K., Schreiber, S., De Santis, A., Okada, Y. and Scheid, P. (1994) Ion-sensitive microelectrode system with short response time. *J. Neurosci. Methods* **51**, 147–153.

Oesch, U., Ammann, D. and Simon, W. (1987) Cell contamination due to the use of carrier-based microelectrodes. *Can. J. Physiol. Pharmacol.* **65**, 885–888.

Solsky, R. L. (1988) Ion-selective electrodes. *Anal. Chem.* **60**, 106R–112R.

Schweitzer, J. S. and Williamson, A. (1995) Relationship between synaptic activity and prolonged field bursts in the dentate gyrus of the rat hippocampal slice. *J. Neurophysiol.* **74**, 1947–1952.

CHAPTER 14

FLUORESCENT DYES FOR MEASUREMENT OF INTRACELLULAR ION CONCENTRATIONS

N. K. Mahanty
Department of Human Physiology, Medical Sciences, University of Newcastle

INTRODUCTION

Standard electrophysiology techniques use a patch- or micro-electrode to measure the current flowing across the cell membrane. It would be useful to know which ions flow to generate the transmembrane current, where they enter the cell, and what local concentrations are attained during ion flow. It is possible to determine the types of ions underlying the current by examining the reversal potential of the current or by ion substitution.

Spatial information on ion distribution can be obtained by using an ion sensitive fluorescent dye, usually in conjunction with standard electrophysiology techniques. A dye fluoresces when it is illuminated with light which can excite the dye molecule to a higher energy state. When the molecule relaxes back to its base energy state it releases the excess energy as a photon. The range of wavelengths which will excite a molecule and cause it to fluoresce is called the excitation spectrum. The released photon always has a longer wavelength than the exciting light. The range of wavelengths, or colors, which a molecule emits when it is illuminated with a specific excitation wavelength is called its emission spectrum. An ion-sensitive fluorescent dye is a dye whose excitation or emission spectra changes when it binds a particular ion. By measuring the change in fluorescence one can calculate the concentration of the free ion in the cytoplasm. This chapter will discuss calcium sensitive fluorescent indicators since calcium is the most commonly studied ion, although dyes are available for a variety of ions including Na^+, K^+, Cl^-, and H^+.

EXPERIMENTAL OVERVIEW

During a typical experiment a cell, in a brain slice or primary culture, is injected with an ion-sensitive fluorescent dye and visualized on a high power microscope. The image of the cell is digitized and stored on a computer. The cell is illuminated with light of the appropriate wavelength to excite the dye and the emitted fluorescence from the cell captured to disk or videotape. This exposure is repeated while the cell is subjected to the experimental conditions, such as addition of a ligand or a series of depolarising pulses. Afterwards, the image data is analysed in order to extract information on the (spatial) distribution, time course, and magnitude of the ion flow during the experiment.

EQUIPMENT

Imaging systems are usually built around an existing microscope. Most systems are based on a transmission microscope where the specimen sits between the light source and the objective. That is, the light is *transmitted* through the specimen into the objective. An inverted or epifluorescence microscope illuminates the specimen through the objective and captures *reflected* light from the specimen back through the objective to the camera. A schematic of a transmission fluorescence system is shown in Figure 14.1.

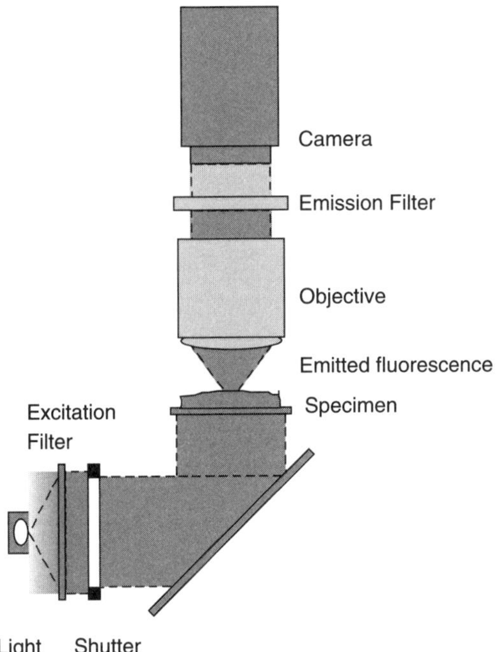

Figure 14.1 A schematic diagram showing the main elements of a upright transmission fluorescence microscope. Light from the lamp is filtered by the band-pass filter before passing through a shutter. The beam is reflected up through the specimen, and the emitted light is collected by the objective lens. A long pass emission filter separates the fluorescent wavelengths from the beam. The fluorescence is detected by the camera.

Light Source

The three most common light sources for fluorescence microscopy are a xenon (Xe) arc bulb, mercury (Hg) arc bulb or a laser. Lasers are usually used only on confocal scanning laser microscopes (CSLM). They have the advantage of providing intense even illumination. Their main drawback is that lasers typically produce only one wavelength of light which limits which dyes can be used. Multiple wavelength lasers and/or ultraviolet lasers are expensive and very uncommon. Hg lamps have good UV emission but have a very uneven intensity spectra, which makes it difficult to get two similar intensity beams at different wavelengths. Xe lamps have less UV output but have a more even intensity spectrum. They are the most commonly used source. There are several points to remember when using arc lamps. The bulb, if dirty, may heat unevenly and explode. (Never handle the bulb with bare hands and avoid getting any dirt or grease on the bulb.) Also, the arc intensity can vary over time. This can be minimized by avoiding turning the lamp on and off unnecessarily and ensuring that it has warmed up sufficiently before beginning experiments.

Excitation filter/shutter

The illuminating light must be filtered to a narrow bandwidth centred on the excitation frequency of the dye. Colored glass filters do not have a narrow enough bandwidth and are not used for these applications. The remaining choices are a monochromator or an interference filter. Monochromators use a movable diffraction grating or prism to separate out the spectrum. By moving the grating different wavelengths of light can be selected. Mononchromators are expensive and although their performance and ease of use make them a better choice than filters, they are not used in most entry level systems. Interference filters are designed to cause all wavelengths, apart from the desired wavelength, to destructively interfere within the filter. Interference filters with a narrow transmission of 5–10 nm are the usual choice for most imaging systems. Transmission drops off with the incidence angle of the excitation beam, so to obtain maximum transmission the filter is placed perpendicular to the beam. Sometimes deliberate tilting of the filter is used to attenuate the light and to alter the transmitted wavelength. Interference filters can be destroyed by temperatures above 70°C. Thus the coated front surface must be oriented towards the lamp since the front coating contains a reflective coating which rejects the unwanted heat radiation. Different filters may be rotated into the light path using a filter wheel in order to select different wavelengths.

A shutter is used to minimize the exposure of the preparation to the light. The shutter may be combined with a filter wheel, such that the different filters or a blank disc can be rotated into the light path. The speed of the filter change and the shutter closure limits how quickly the system can switch between two wavelengths. The speed with which a system can switch filters usually limits how quickly images can be acquired and hence sets the limit on the temporal resolution of the data. That is, if it takes 30 ms to change filters than it will be impossible to detect transient changes in ion concentration with a time resolution less than 30 ms.

Microscope

The specimen is usually visualized with a fluorescence microscope. The microscope optics, comprising of objective, condenser, barrier filters, mirrors, and guiding lenses must be able to pass the wavelengths of interest with minimum attenuation. Fluorescence microscopes are usually optimized to have the least number of elements in the optical path. Microscopes can pass light in the 'visible' spectrum from short wavelength ultraviolet (330 nm) up to the long wavelength infrared (900 nm). Glass attenuates UV dramatically and most microscopes designed for using short wavelength dyes in the 340–390 nm range, such as Fura-2, INDO, and BCEF, use special quartz optics. The emission filter in front of the camera aperture is usually a long pass filter designed to only pass the fluorescent light. The difference between the excitation wavelength and the emission wavelength is called the Stokes shift. The bigger this number, the more easy it is to separate the emitted light from the exciting beam.

Detectors (Cameras)

The fluorescence signal is captured by a camera and either stored on videotape or digitally on a computer. If it is stored on videotape during the experiment, then it is digitized later and analysed on the computer. Depending on the speed of the computer all the images may be analysed and displayed during the experiment (on-line) or later (off-line). There are three main types of detectors: Photomultipliers (PMTs), Charge Coupled Devices (CCDs), or Silicon Intensified Tube (SIT) cameras.

Photomultipliers

Photomultipliers are similar to a vacuum tube. A photon hits the coated front surface dislodging an electron which is accelerated through a series of dynodes liberating more electrons as it passes. A current is produced at the anode of the device proportional to the number of photons hitting the front surface. PMTs are extremely sensitive and can detect single photons. But, they don't preserve any spatial information since a photon hitting anywhere on the front surface will liberate electrons from the entire surface of the next stage. PMTs are mainly used in CSLMs where the image is built up sequentially. They can be damaged irreversibly by intense light and care must be taken not to over expose them.

SIT cameras

SIT cameras and other image intensifiers use the same architecture as a photomultiplier but constrain the photons by using a microchannel plate. The photons cannot wander all over the intermediate dynodes but are constrained by a microchannel tube to the same spatial location. (Rather like shining light down a set of closely packed straws.) The electrons coming out of the intensifier hit a phosphorescent screen and the image is detected by a standard video camera.

Cooled CCD cameras

Cooled CCD cameras are mainly used for high spatial resolution experiments. They are simply an array of solid state photodiodes fabricated on a single silicon integrated circuit (chip). They produce electrons when a photon strikes the front surface. Unlike a PMT a CCD camera cannot resolve single photon events from the background noise. CCDs have a higher quantum efficiency than photomultipliers and can detect low light fluorescence while preserving the spatial information in the image. By averaging images very low light level images can be obtained. They have a linear response across their entire imaging area, are the most sensitive cameras available and the least prone to damage by intense light sources. Good cameras are expensive and usually produce a non standard video signal which need some specialized electronics to display the image.

PROBLEMS/CAVEATS

There are a number of technical problems associated with using fluorescent dyes. They all stem from one major assumption concerning the fluorescent data. That is, all the measured fluorescence, and changes in fluorescence, are related to changes in concentrations of the ion. Thus, if two regions of the cell give different fluorescence measurements, we would assume that there are different concentrations of the ion present in those regions. If other factors affect the detected fluorescence intensity then they may affect the validity of the data. Factors which can affect the measured fluorescence include:

Autofluorescence

Certain organic molecules can autofluoresce when illuminated with certain wavelengths of light and this fluorescence can contribute to the measured signal. The solution is to lower the intensity of illumination until the autofluorescence is not detectable or to choose a wavelength where autofluorescence is at a minimum.

Non-uniform dye distribution

If the dye concentration is non-uniform throughout the cell, the regions with more dye will yield a higher intensity signal. The solution to this problem is to allow the dye to diffuse throughout the cell for a sufficiently long time so as to ensure an equal dye concentration, or to use a 'ratiometric' dye.

Photobleaching/phototoxicity

Most dyes breakdown (photobleaching) with repeated exposure to light and the by-products can be toxic to the cell. Exposure to light should be minimized and the minimal excitation intensity necessary to provide a good image should be used. Alternatively, one can add antioxidants or reduce oxygen in the preparation, but this can interfere with most live preparations.

Different cell thickness and optical path length

The emitted fluorescence from distinct regions of the cell can be attenuated differently by varying slice thicknesses and optical path lengths, resulting in signals from thick parts of the tissue being more attenuated than signals with a shorter path length. The solution is to use a ratiometric dye or to examine small regions of the specimen which have comparable thicknesses.

Uneven illumination

If the intensity distribution of the light source across the cell is uneven, certain regions which receive more light will appear brighter. The solution is to adjust the optics to give a uniform illumination or to use a ratiometric dye.

Uneven detector sensitivity

The detector used to image the fluorescence may have a non-uniform sensitivity across the face of the detector. Most video tube cameras are more sensitive in the central region compared with the edges. The solution is to correct the image for sensitivity differences, use a linear detector such as a CCD camera, or to use a ratioable dye.

Dye quenching

Heavy metals such as cadmium can bind to the dyes and quench the fluorescent signal resulting in a lower fluorescent signal. The solution is to carefully screen what experimental solutions are used for the experiment.

Changes in K_D

Calculating ion concentrations from the fluorescent data requires that the dissociation constant (K_D) of the dye be known. Increases in viscosity and temperature can reduce the K_D of certain indicators.

Signal buffering

The dye binds ions, and may be such a strong buffer that it disrupts the ion sensitive cellular processes under investigation. The solution is to use the lowest concentration of dye necessary.

Dye saturation

It is possible for the dye to become saturated with the ion so that subsequent increases in ion concentration are not accompanied by an increase in fluorescence. The solution is to use a dye whose K_D is about one-half/two-thirds the expected peak concentration.

TECHNICAL CONSIDERATIONS

Despite all these caveats it is possible to use ion sensitive dyes to produce meaningful data on ion concentrations during biological events. The following sections will cover the techniques involved in loading, imaging and calibrating two popular calcium indicators: Fluo-3 and Fura-2.

Dye loading

There are two ways of filling a cell with dye. A membrane impermeable free acid, or ionic salt, form of the dye can be injected into the cell through a patch- or micro-electrode. This loads one cell at a time and if loaded from a patch electrode yields a rough estimate of the concentration of the dye in the cell (i.e. the same as the concentration in the electrode).

Alternatively, one can use a membrane permeable form of the dye. Usually this is the impermeant dye molecule with four lipophilic acetoxy-methylester groups bound to it. Once inside the cell cytosolic esterases cleave the ester bonds yielding the impermeant form of the dye. There are several problems with this technique:

- *Partial desterification:* If all four acetoxy methyl ester groups are not removed the dye will still fluoresce but it will not be sensitive to the calcium concentration.
- *Organelle sequestration*: The dye may pass through the cytosol into organelles such as endoplasmic reticulum or mitochondria before the esters are cleaved. Their fluorescence will not reflect change in cytosolic calcium.
- *Active pumping*: Certain cells have non-specific anionic channels which can extrude the negatively charged fluorescent dye. For example, in rat hepatocytes which were loaded with 5 μM Fura-2 approximately 83% of the dye was localized to the cytosol, 10% to the organelles, and 6% to the mitochondria (Roe et al., 1990). These numbers may vary for different cell types.

The solution to all these problems is to be careful with loading times and conditions. Pre-chilling the cells to 5°C and then loading at 15–37°C for 15 minutes works well. Anionic channel inhibitors can be added to help prevent the extrusion of the dye.

Fluo-3

Fluo-3 is a commonly used calcium dye based on BAPTA. It has a K_D of 400 nM and when excited by light in the 440–500 nm range emits fluorescence around 530 nm (Figure 14.2). The fluorescent signal increases with free calcium concentration and there is an overall 12-fold change in intensity between zero calcium and a saturated signal of about 40 mM calcium. There is some fluorescence at zero calcium, meaning that the cell can be seen at resting levels of calcium. Fluo-3 is available in both acid and ester forms. Usually the dye is injected through a patch electrode at 100–200 mM concentrations.

Fura-2

Fura-2 is a ratiometric dye. Its excitation peak wavelength shifts from 340 to 380 nm as calcium decreases. This can be seen in Figure 14.3. At high free calcium concentrations the fluorescence emission when exciting at 340 nm is large. For the same calcium concentration the fluorescence at 380 nm is very low. Thus when illuminated at these two wavelengths and captured at 560 nm the ratio of the two signals will give a measure of calcium which is independent of variations in dye loading, local optical path length, non-uniform illumination and non-uniform detector sensitivity. This is because the two

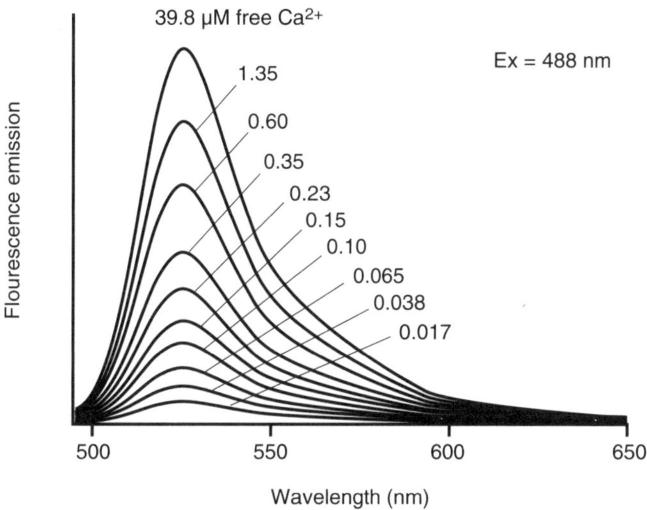

Figure 14.2 Emission spectra of pentapotassium Fluo-3 at room temperature for ten different calcium solutions. Concentration of free calcium ranges from 0.017–39.8 µM. The dye was excited at 488 nm and emission measured between 500 and 650 nm. Note that fluorescence increases with free calcium concentration. (Figure reproduced with permission from Molecular Probes.)

images at each point have exactly the same optical characteristics and it is the ratio of the intensity, not the value of the intensity, which gives the measure. This ratiometric technique works because for one wavelength fluorescence increases with free calcium and for the other wavelength fluorescence decrease as calcium increases.

The K_D for Fura-2 is 250 nM meaning that it will saturate at lower calcium concentrations then Fluo-3. The fluorescence obtained while measuring high calcium levels with 380 nm light is a very small signal. So comparing two measures of high calcium will result in the comparison of two small numbers each of which can have a high error since the signal to noise ratio will be low. Comparing two small numbers can result in measurement errors. An alternative is to measure at 360 nm, which is the intersection point of the curves (isoemissive or isobestic point) and is more easily measured. Both 340 and 380 nm are in the ultraviolet range of light, so the microscope optics must be able to transmit UV light with minimum attenuation.

Calibration

Try to avoid calibrating fluorescent dyes. Calibrating the dye is time consuming, difficult, and prone to error. For most applications it is not necessary since most experimenters are interested in relative changes of calcium not absolute values. Data can simply be represented as a change in intensity relative to a baseline image. If absolute values must be known there are two different calibration techniques.

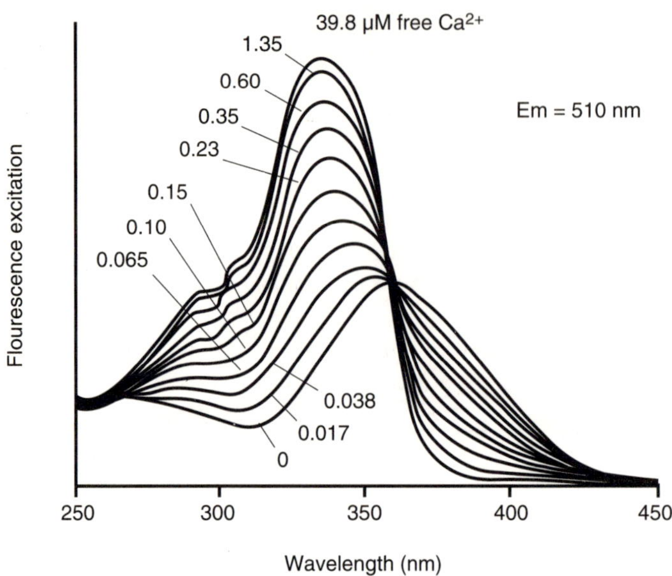

Figure 14.3 Excitation spectra of pentapotassium Fura-2 at room temperature for ten different calcium solutions. Concentration of free calcium ranges from 0.0–39.8 μM. Fluorescence was measured at 510 nm while exciting between 250 and 450 nm. (Figure reproduced with permission from Molecular Probes.)

Calibration *in vitro*

Calibration *in vitro*, is where a solution containing the dye and various amounts of calcium and EGTA is measured in a fluorometer or spectrophotometer. EGTA binds calcium with a known K_D, allowing the amount of free calcium in solution to accurately calculated. The amount of free calcium is calculated and the equivalent fluorescence level noted. These levels are then equated to the measured fluorescence in the preparation. There are problems with this technique. The K_D for calcium may be different in solution compared with the cell. The amount of free calcium can differ with viscosity and temperature. The cuvette and solution may not approximate the cell size and optical path length. The fluorometer sensitivity may be different than that of the detector used in the experiments.

Calibration *in vivo*

After the experiment the cell is permeablized with a calcium ionophore such as ionomycin. Calcium enters the cell and saturates the dye. This maximum fluorescence (Fmax) is measured at two wavelengths and the ratio (Rmax) is calculated. Then a heavy metal, usually manganese chloride, is added to the bath to quench the fluorescent signal in order to measure the minimum fluorescence (Fmin) at both wavelengths. Alternatively, Fmin can be measured, and hence the minimum ratio (Rmin) calculated, by replacing the external saline with zero calcium solution. For a ratiometric dye the free calcium concentration ([Ca]i) can be calculated from Equation 1 (Grynkiewicz et al., 1985):

$$[Ca]_i = K_D \,.\, B.\, (R - Rmin)/(Rmax - R) \quad (1)$$

where K_D is the dissociation constant for the dye, B is a scaling factor given by the ratio of the fluorescence at one wavelength in conditions of zero calcium and saturating calcium, R is the fluorescence ratio of the two wavelengths, Rmin is the fluorescence ratio of the two wavelengths for zero calcium conditions, and Rmax is the fluorescence ratio at the two wavelengths for a saturating level of calcium.

For a non-ratiometric dye such as Fluo-3, formula 2 is used to calculate the free calcium concentration. The equations are similar, with fluorescence level (F) substituted for the ratio value R:

$$[Ca]_i = K_D \,.\, (F - Fmin)/(Fmax - F) \quad (2)$$

where F, Fmin and Fmax are the actual fluorescent intensity values.

There are problems with calculating calcium concentration. Ionomycin is pH sensitive and efficacy peaks at pH 9.5 and drops off below 7.0. (An alternative ionophore is 4BromoA23187, or a mild detergent such as digitonin.) If the dye leaks out through the pores during the calibration process the calibration will be skewed. Calcium entry through the pores can activate the Ca:ATP pump, as well as the Na/Ca exchanger and internal calcium pumps. So the measured calcium level may be changing over time.

RECENT PAPER

A demonstration of the strength of calcium imaging is given in a paper by Muller and Connor (1991). An important question in neuroscience is: do dendritic spines act as a diffusional barrier for calcium? This is important because various learning and memory models require some form of cellular diffusional barrier. Muller and Connor injected Fura-2 from a microelectrode into single pyramidal CA1 neurons in a hippocampal brain slice. By gently stimulating the afferent fibers and using a cooled CCD imaging system and home-grown software, they obtained the first pictures of calcium entry into a dendritic spine and showed that levels in the spine could be elevated higher than those in the adjacent dendrite, proving that the spine neck can act as a diffusional barrier.

REFERENCES AND FURTHER READING

Cell Calcium (1990) **11**, Entire issue.

Grynkiewicz, G., Poenie, M. and Tsien, R. Y. (1985) A new generation of calcium indicators with greatly improved fluorescent properties. *J. Biol. Chem.* **260**, 3440–3450.

Haugland, R. P. (ed) (1996) *Handbook of Fluorescent Probes and Research Chemicals*. Oregon: Molecular Probes.

Mackenzie, P. J., Umemiya, M. and Murphy, T. H. (1996) Ca^{2+} imaging of CNS axons in culture indicates reliable coupling between single action potentials and distal functional release sites. *Neuron* **16**, 783–795.

Moore, E. D., Becker, P. L., Fogarty, K. E, Williams, D. A. and Fay, F. S. (1990) Ca^{2+} imaging in single living cells: theoretical and practical issues. *Cell Calcium* **11**, 157–179.

Muller, W. and Connor, J. A. (1991) Dendritic spines as individual neuronal compartments for synaptic Ca^{2+} responses. *Nature* **354**, 73–76.

Ogden, D. (ed) (1994) *Microelectrode Techniques. The Plymouth Workshop Handbook* (2nd edn). Cambridge: The Company of Biologists Ltd.

Roe, M. W., Lemasters, J. J. and Herman, B. (1990) Assessment of Fura-2 for measurements of cystolic free calcium. *Cell Calcium* **11**, 63–73.

Tsien, R. Y. (1981) A non-disruptive technique for loading calcium buffers and indicators into cells. *Nature* **290**, 527–528.

Tsien, R. Y. and Harootunian, A. T. (1990) Practical design criteria for a dynamic ratio imaging system. *Cell calcium* **11**, 93–109.

Yuste, R. and Tank, D. W. (1996) Dendritic integration in mammalian neurons, a century after Cajal. *Neuron* **16**, 701–716.

SECTION 6

IN VIVO TECHNIQUES AND PREPARATIONS

CHAPTER 15

STEREOTAXIC PLACEMENT OF PROBES IN NEUROBIOLOGY

S. P. Perrett
Department of Neurobiology
Civitan International Research Center, Birmingham

INTRODUCTION

Many neurobiologists are interested in how different areas of the brain interact to produce specific behaviors. However, such interactions are extremely complicated, even for simple behaviors, and, thus, discrete areas of the brain are usually tested individually for their involvement in a particular behavior. In this way it can first be determined that an area contributes to the behavior and then experiments can be designed to assess the nature of that contribution. Once the contributions of different sites to a behavior have been determined individually, the way that these sites interact to produce the behavior can then begin to be inferred.

An indispensable technique in experiments of this type is the placement of probes in regions of the brain thought to be involved in a behavior. These probes are then used to perturb normal functioning or to assay activity in that area of the brain during the behavior. Stereotaxic techniques are used to position such probes within discrete regions of the brain in intact animals with some degree of consistency. In this chapter I will explain how stereotaxic techniques are used to position probes and ways in which lesions made with such probes can be used to gain insight into the neural pathways and mechanisms involved in simple behaviors.

WHAT ARE STEREOTAXIC TECHNIQUES?

A starting point for understanding what is meant by stereotaxic techniques is to define the roots of the word. Stereo literally means three-dimensional, while taxic means to arrange. Thus, stereotaxic techniques are used to arrange or position objects in three-dimensional space. In neurobiological applications the objects are often electrodes or cannulae which are to be positioned within discrete regions of the brain. The aim of stereotaxy is to position these objects in precisely the same place in different animals while minimizing damage to overlying and adjacent tissue.

Two instruments are needed for stereotaxic placements (Figure 15.1). First, a headholder must be used which can both constrain the movement of the animal's head and also position the head in a consistent orientation across animals. The headholder must also have mounted a bar to which the stereotaxic device itself can be affixed. Such headholders are commercially available for certain species or can be custom built. Second, the stereotaxic device is used to actually position the probe to be used. This device most commonly consists of a clamp to hold the probe which is attached to an arm which can be moved independently in three planes. This is usually accomplished through a series of screws which can move the arm along tracks either up and down, side-to-side, or forwards and backwards. The amount of movement in each plane can be measured on scales which have been built into the stereotaxic manipulator. These scales are precisely calibrated to allow accurate measurement of millimeters of movement. A number of different types of stereotaxic manipulators designed for use on different animals are commercially available.

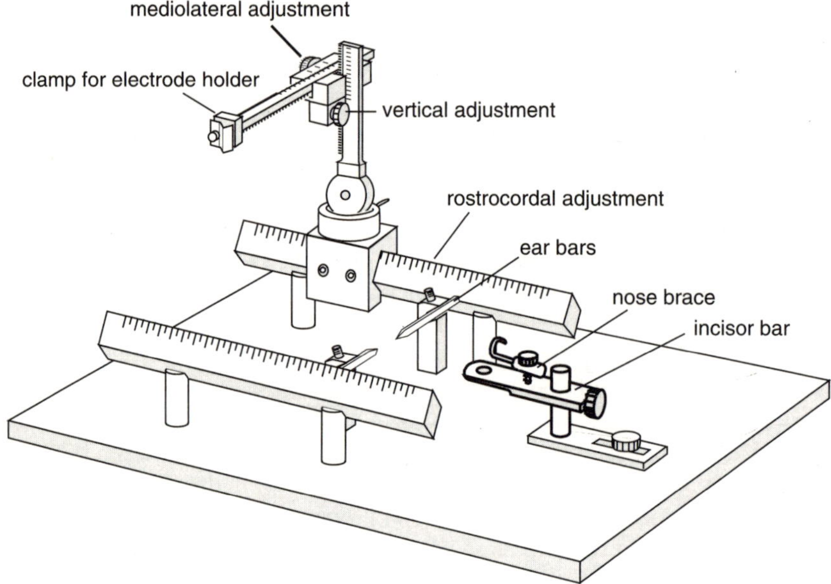

Figure 15.1 Headholder for the rat with an attached stereotaxic device. The animal's head is maintained in the appropriate orientation by ear, nose, and incisor bars. A screw-down clamp on the probe carrier keeps the probe to be inserted in the correct position. The probe can be moved independently in three planes with knobs on the stereotaxic device. The amount of movement in each plane can be precisely measured on scales built into the stereotaxic device and consistent placement of probes in different animals can be accomplished by referencing this movement to landmarks on the animal's skull.

To place a probe within the brain of an animal requires a stereotaxic co-ordinate for each of the three planes of movement. Many such sets of co-ordinates have been established by researchers for specific nuclei and areas of the brain in a number of popular experimental animals. Also, for most popular research animals, stereotaxic atlases exist from which co-ordinates can be taken (e.g. Paxinos & Watson, 1986). These atlases often display the whole brain of a specific animal which has been sectioned in different ways and for which a stereotaxic scale is superimposed on each section. Stereotaxic co-ordinates are generally derived in reference to sutures on the animal's skull, which are relatively consistent across animals, when the animal's head is in a specific orientation. If co-ordinates have not been established for a particular site and no atlas exists to provide a starting set of co-ordinates, then they may have to be derived for a particular experiment by trial and error (e.g. Schuller et al., 1986; Ireland & MacLeod, 1993).

USING PROBES POSITIONED WITH STEREOTAXIC CO-ORDINATES TO MAKE LESIONS

A number of different types of probes are used for neurobiological applications (e.g. Jonsson, 1981; Bures et al., 1983). Commonly, such probes are used to block or alter neuronal activity, to stimulate neuronal firing, or to assay in some way neuronal activity in a discrete region of the brain. Thus, experiments using such probes could loosely be classified as lesion experiments, stimulation experiments, or recording experiments, respectively. In this section I will discuss how discrete lesions can be used to identify pathways involved in simple forms of learning.

The first step in determining the neurobiological basis of a behavior is to determine which sites in the brain are involved. One approach is to remove or inactivate an area of the brain thought to be important for the behavior and assay how this pertubation affects the behavior. In the extreme, relatively large areas of the brain can be removed or isolated from the rest of the brain to determine if these areas are needed for the behavior. These type of experiments are usually done to exclude the involvement of large areas of the brain. For example, it can be shown that the brainstem and cerebellum are sufficient to produce a learned motor reflex because a cut through the brain at the level of the thalamus does not affect that reflex (Mauk & Thompson, 1987). However, to determine the precise neural pathways involved in a behavior and begin to ascertain the nature of the involvement of specific sites, very discrete lesions are desirable.

Discrete lesions are commonly made either electrolytically or pharmacologically (for examples of other lesion techniques see Jonsson, 1981; Bures et al., 1983). Each technique requires the positioning of probes using stereotaxic co-ordinates. Electrolytic lesions involve passing sufficient direct current through chronically implanted electrodes to cause the death of neurons in the immediate vicinity of the electrodes. If the lesion disrupts the behavior, it can be assumed that the lesioned area was contributing to the behavior in some way. By systematically making lesions in different sites, often in tandem with stimulation of these sites through electrodes, critical neural pathways for simple behaviors can be revealed. For an elegant example of the use of these techniques, refer to studies done by Davis and colleagues in examining the acoustic startle response in rats (e.g. Davis et al., 1982; Berg & Davis, 1985; Lee et al., 1996).

The advantage of electrolytic lesions is their simplicity. The electrodes can be inserted, anchored to the skull, and protected from dislodgement easily. At the appropriate time the lesion can be made simply by attaching a stimulator to leads connected to the electrodes and passing current for several minutes. The lesion extent can also be established to some degree using rather simple histological techniques. The disadvantage of this technique is that the lesions are permanent. Long-term effects of permanent lesions, such as sprouting of neurites or changes in receptor sensitivity of viable neurons remaining in the area of the lesion, may mask the function of the destroyed tissue. For these reasons electrolytic lesions are usually done as a quick and easy way to determine initially if an area is part of a pathway involved in the behavior.

Much more information about the nature of an area's involvement in a behavior can be obtained when neuronal activity can be blocked or altered transiently. For this reason, techniques for making 'reversible' lesions have been developed and refined. A popular way to alter or block neuronal activity transiently is by infusion of pharmacological agents through chronically implanted cannula. With pharmacological lesions, the agent to be infused depends on the rationale for that particular experiment. For example, lidocaine, which blocks sodium channels, may be infused to block all activity in an area, including that of neurons and fibers of passage. However, muscimol, a GABA receptor agonist, may be infused to block only neurons in that area, while sparing fibers of passage. If the experimenter wishes to block inhibition in an area then a GABA receptor antagonist such as bicuculline may be used. An advantage to this technique is that lesions may be restricted not only to a particular area but also to blocking the activity of particular molecules.

There are other advantages of reversible lesions over permanent lesions when examining learned behaviors. With reversible lesions the effect of the lesion at different times during learning or on different aspects of the same behavioral paradigm can be examined in the same animal. For example, it might be that an area of the brain is more important early in learning than after the animal is well-trained. With reversible lesions this can be shown in individual animals by infusing agents at different times during training. Similarly, an area may be important for learning a conditioned response but not extinguishing that response. Again, this can be tested in the same animal with reversible lesions. It is more informative to test such possibilities in individual animals because lesions in different animals are never identical. Also, fewer animals may be needed to obtain the same data with reversible lesions.

Perhaps, the greatest advantage in the use of reversible lesions is the potential to identify sites where the learning occurs (Chapman et al., 1990; Krupa et al., 1993). This arises from the capability of dissociating learning and expression of simple behaviors in the same animal. This is done by inactivating the site of interest while training occurs and then assessing learning after recovery from the infusion. If learning occurs during inactivation then the inactivated site must be concerned with the output or expression of the behavior. If learning has not occurred during the inactivation, then the site that has been inactivated may be the site of learning or a critical afferent to the site of learning.

Further, mechanisms for the way in which neurons implement learning may begin to be revealed by infusion of pharmacological agents. For example, Davis and colleagues (Campeau et al., 1992) have shown that infusion of the NMDA receptor antagonist, AP5, into the amygdala, an area suspected of being the site of learning for a fear-potentiated acoustic startle response, blocks learning but does not affect the expression of already learned fear-potentiated startle. This suggests that the learned fear which potentiates acoustic startle is dependent on the activation of NMDA receptors in the amygdala. However, if fear-potentiation of the startle response has already been learned, it can still be expressed despite the block of NMDA receptors. Expression is mediated through AMPA receptor activation (Kim et al., 1993).

Problems encountered with the infusion technique for making lesions is that more hardware must be exposed on the animals' heads. This makes it more likely that a cannula will be broken or bent before the required data is acquired. Also, much more complicated and less reliable techniques must be used to approximate the extent of the lesion with infusion (Sandkuhler et al., 1987). This is because the extent of diffusion of the effective concentration of the pharmacological agent is the variable to be determined. Controls utilizing infusion of vehicle (such as saline) should also be done to ensure that the observed effect is due to the active agent. Despite these difficulties, the value of data that can be obtained with reversible lesions generally makes this technique superior to electrolytic lesions.

These examples show how the stereotaxic positioning of probes can be used to determine neural pathways involved in simple learned behaviors. By combining a lesion and stimulating approach, the pathways and the sites of learning can be determined and even mechanisms for learning can begin to be identified. Once this has been done the mechanisms for learning a defined behavior can be examined more closely, both in the intact animal and in dissociated preparations. Examining these mechanisms in the intact animal may be done with recording experiments which also require stereotaxic positioning of probes. Examples of such experiments include recording electrophysiological activity and dialysis of extracellular fluids to assay biochemical changes during learning. Even as new technologies develop that allow finer resolution of cellular activity in intact, behaving animals, stereotaxic placement of probes will undoubtedly still be required.

RECENT PAPERS

Stereotaxic placements of stimulating and recording electrodes are used in two recent experiments done in the lab of LeDoux and colleagues (Rogan & LeDoux, 1995; Quirk et al., 1995) which examine the effect of long-term potentiation (LTP) and fear conditioning on auditory responses of neurons in the lateral nucleus of the amygdala (LA). The LA is a promising candidate for a site of plasticity underlying fear conditioning. In the first paper (Rogan & LeDoux, 1995), LTP of synaptic strength in the LA is induced by high frequency stimulation in the medial geniculate body. It is demonstrated that the response to a naturally transduced acoustic stimulus is also potentiated by this high frequency stimulation. In the second paper, Quirk et al. (1995) show that fear conditioning also can increase the magnitude of tone-elicited responses in the LA. Taken together, these experiments provide important evidence that artificially enhanced synaptic responsiveness and behavioral learning may share common mechanisms.

REFERENCES AND FURTHER READING

Berg, W. K. and Davis, M. (1985) Associative learning modifies startle reflexes at the lateral lemniscus. *Behav. Neurosci.* **99**, 191–199.

Bures, J., Buresova, O. and Huston, J. P. (1983) *Techniques and Basic Experiments for the Study of Brain and Behavior*. Amsterdam: Elsevier.

Campeau, S., Miserendino, M. J. D. and Davis, M. (1992) Intra-amygdala infusion of the N-methyl-D-aspartate receptor antagonist AP5 blocks acquisition but not expression of fear-potentiated startle to an auditory conditioned stimulus. *Behav. Neurosci.* **106**, 569–574.

Chapman, P. F., Steinmetz, J. E., Sears, L. L. and Thompson, R. F. (1990) Effects of lidocaine injection in the interpositus nucleus and red nucleus on conditioned behavioral and neuronal responses. *Brain Res.* **537**, 149–156.

Davis, M., Gendelman, D. S., Tischler, M. and Gendelman, P. M. (1982) A primary acoustic startle circuit: lesion and stimulation studies. *J. Neurosci.* **2**, 791–805.

Ireland, W. P. and MacLeod, W. J. (1993) A method for finding stereotaxic co-ordinates from brain sections. *J. Neurosci. Meths.* **49**, 93–96.

Jonsson, G. (1981) Lesion Methods in Neurobiology. In *Techniques in Neuroanatomical Research*, edited by C. H. Heym and W. -G. Forssmann, pp. 71–99. Heidelberg: Springer.

Kim, M., Campeau, S., Falls, W. A. and Davis, M. (1993) Infusion of the non-NMDA receptor antagonist CNQX into the amygdala blocks the expression of fear-potentiated startle. *Behav. Neural Biol.* **59**, 5–8.

Krupa, D. J., Thompson, J. K. and Thompson, R. F. (1993) Localization of a memory trace in the mammalian brain. *Science* **260**, 989–991.

Lee, Y., Lopez, D. E., Meloni, E. G. and Davis, M. (1996) A primary acoustic startle pathway: Obligatory role of cochlear root neurons and the nucleus reticularis pontis caudalis. *J. Neurosci.* **16**, 3775–3789.

Mauk, M. D. and Thompson, R. F. (1987) Retention of classically conditioned eyelid responses following acute decerebration. *Brain Res.* **403**, 89–95.

Paxinos, G. and Watson, C. (1986) *The Rat Brain in Stereotaxic Coordinates*. Sydney: Academic Press.

Quirk, G. J., Repa, J. C. and LeDoux, J. E. (1995) Fear conditioning enhances short-latency auditory responses of lateral amygdala neurons: Parallel recordings in the freely moving rat. *Neuron* **15**, 1029–1039.

Rogan, M. T. and LeDoux, J. E. (1995) LTP is accompanied by commensurate enhancement of auditory-evoked responses in a fear conditioning circuit. *Neuron* **15**, 127–136.

Sandkuhler, J., Maisch, B. and Zimmermann, M. (1987) The use of local anaesthetic microinjections to identify central pathways: A quantitative evaluation of the time course and extent of the neuronal block. *Exp. Brain Res.* **68**, 168–178.

Schuller, G., Radtke-Schuller, S. and Betz, M. (1986) A stereotaxic method for small animals using experimentally determined reference profiles. *J. Neurosci. Meths.* **18**, 339–350.

CHAPTER 16

IN VIVO MICRODIALYSIS

A. J. Lawrence
Department of Pharmacology, Monash University, Melbourne

INTRODUCTION

Neurotransmission, defined as the propagation of nervous information, can be studied in a number of ways. Electrophysiological techniques can characterize the physical properties of neurons, e.g. conduction velocity, frequency of firing, while neurochemical techniques allow unequivocal chemical identification of the neurotransmitter molecules. One such method utilized to measure the release of endogenous neurotransmitters is *in vivo* microdialysis. The technique of *in vivo* microdialysis was designed as a refinement of more limited methods to measure transmitter release such as cortical cups and push-pull cannulae, and has become established as a widely used tool in experimental neuroscience. The aims of this chapter are to explain (i) the theoretical basis of microdialysis; (ii) the practical approach, including construction of a dialysis probe; (iii) the applicability of the technique; and (iv) the shortcomings of the technique.

PRINCIPLES OF *IN VIVO* MICRODIALYSIS

The principles of intracerebral microdialysis have been thoroughly reviewed (e.g. Benveniste & Hettemeier, 1990; Justice, 1993). As indicated in the name of the technique, the underlying principle of *in vivo* microdialysis is dialysis, or diffusion of soluble molecules across a semi-permeable membrane. A schematic diagram of a microdialysis probe in use is shown in Figure 16.1. Following stereotaxic implantation into the brain region of interest, the membrane is constantly perfused with a solution that initially lacks the substance(s) of interest, but that closely mimics the endogenous extracellular fluid both in ionic composition and pH—artificial cerebrospinal fluid (aCSF) is often used. Diffusion of the endogenous extracellular fluid and its dissolved neurochemicals occurs across the resultant concentration gradient until steady state equilibrium is regained. The continuous flowing of the perfusate solution through the probe carries the desired samples to an outlet collection point (usually a capillary tube), from where sample analysis can be commenced. Thus, the technique allows sampling of the extracellular compartment of the brain, and therefore samples represent the extracellular overflow of neurotransmitter. Importantly, there is no net loss of fluid from the brain, and low flow rates (typically 1–2 μl/min minimize hydrostatic pressures). Following calibration of the dialysis membrane efficiency, the concentration, within that particular brain region, of the pertinent neurotransmitter can be estimated.

The microdialysis probe can also be utilized as a drug delivery system, allowing

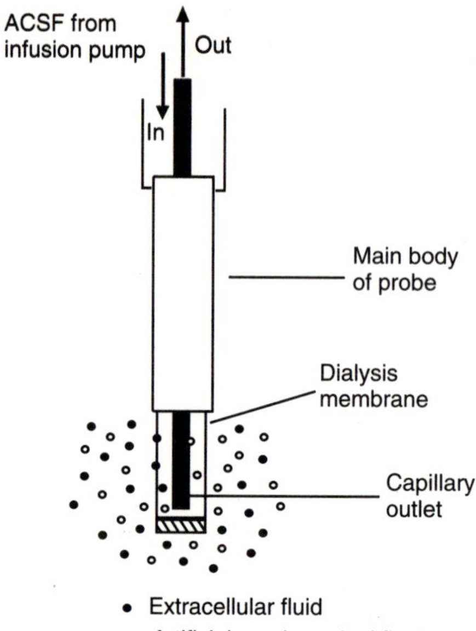

Figure 16.1 Schematic diagram of a microdialysis probe.

specific targeting of drug administration to the area of interest. In this case, the diffusion principle applies in reverse, i.e. from inside the probe to the brain extracellular space. It must be remembered that only a fraction of the drug that is perfused through the probe will actually cross the membrane and access the target site, and as such, relatively high drug concentrations are often used.

CONSTRUCTION OF A MICRODIALYSIS PROBE

Dialysis probes are commercially available, although expensive. With practice, probe making can become a routine procedure. There are three main types of dialysis probe: concentric, U-shaped and transverse. Appropriate membranes can be obtained from a renal dialysis pack, the most commonly used being polyacrylonitrile; however, methyl cellulose and other polymers are also used (Hsiao et al., 1990). Many researchers use 'Hospal' polyacrylonitrile membranes that have an external diameter of around 250–300 μm and a molecular weight cut-off of around 20 000 Daltons. The following description relates to the manufacture of a concentric dialysis probe:

1. Insert a wire into the lumen of an individual length of membrane to act as a support during the initial preparation.
2. Insert the membrane into the end of a length of 23G needle tubing (cannula). Ensure that the membrane is well into the cannula (~5 mm) and then secure it in place with

epoxy resin. [Hint: add glue to membrane and then advance it into the cannula—make sure no glue is left on the membrane that you intend to use as the active probe].
3. Withdraw the wire, trim the membrane and allow epoxy resin to dry.
4. Cut a length (20–30 mm) of nylon or polypropylene tubing to fit tightly over the 23G cannula and make a hole ~10 mm from one end with a 25G needle. Place the tubing over the free end of the 23G steel cannula and secure in place with adhesive (i.e. you now have a length of 23G steel cannula with a membrane on one end and an inlet tube on the other end).
5. Trim the membrane to the final desired length (dependent upon the brain region to be studied) and then insert a fine fused silica glass capillary tubing (from Scientific Glass Engineering) through the open end of the membrane. Work the capillary outlet up through the body of the probe and out of the hole made in the inlet tubing.
6. Seal the end of the membrane, ensuring that the capillary outlet is free, and then finally seal the hole where the capillary exits from the inlet tubing.

Before probes are used for *in vivo* experimental purposes, their ability to recover the desired analyte should be assessed *in vitro*. This is particularly important for home-made probes, as it allows comparison of the performance characteristics of individual probes. One should aim to make probes with similar, reproducible properties. Probe calibration is also required for quantitative microdialysis (Justice, 1993).

In vitro recovery of constructed probes can easily be determined by perfusing the dialysis probe with artificial cerebrospinal fluid (ACSF) and placing the probe into a container of ACSF with a known concentration of analyte. After allowing for equilibration, samples can be taken and analysed and the recovery of the probe is simply the ratio of the concentration of the sample collected from the probe outlet and the known concentration in the standard solution. In reality, the absolute recovery of the probe membrane is not of great importance, as long as the probe allows measurement of the analyte of interest and the relative recoveries of individual probes are similar.

SAMPLE ANALYSIS

The method of analysis obviously depends upon the properties of the particular analyte; however, high performance liquid chromatography (HPLC; see Chapter 22) and radioimmunoassay (RIA) are the two most common. Below is a list of commonly studied neurotansmitters and the usual method of analysis:

1. Catecholamines and indoleamines, e.g. noradrenaline, dopamine, 5-hydroxytryptamine: HPLC with electrochemical detection based on oxidation of the sample at the appropriate redox potential.
2. Amino acids, e.g. glutamate, GABA: HPLC with electrochemical or fluorescence detection of chemically derivatised samples.
3. Neuropeptides, e.g. substance P, neurotensin: RIA.

Clearly, to perform *in vivo* microdialysis successfully, strong analytical skills are necessary to maintain sensitive protocols.

APPLICABILITY OF THE TECHNIQUE

As has just been indicated, a major limiting factor for successful microdialysis experiments is a sensitive and reliable assay for the desired analyte(s). The technique can essentially be utilized to measure extracellular levels of anything that will cross the dialysis membrane, as long as a suitable means of analysis exists. *In vivo* microdialysis can be used both in anaesthetized and also conscious animals. When conscious animals are to be employed the dialysis probe is inserted through a previously implanted guide cannula. Most often, microdialysis is used to measure neurotransmitter release in specific brain nuclei; however, the methodology can be utilized for a wide variety of purposes. Listed below are some examples of the different applications of *in vivo* microdialysis in neuroscience research:

1. Identification of the neurotransmitter used by specific neurons (e.g. Lawrence & Jarrott, 1994).
2. Dual dialysis with two probes in different brain regions, often using one for drug administration to a group of cell bodies and the other for measurement of transmitter release at the corresponding terminal field. This type of approach is often used in the study of drugs of abuse and the 'reward' pathway and performed in conscious animals to allow parallel behavioural observations (e.g. Chen et al., 1996).
3. Measurement of extracellular levels of a second messenger (e.g. cyclic GMP) to give an indication of receptor activation and subsequent effector mechanisms (e.g. Fedele & Raiteri, 1996).
4. Concomitant microdialysis of a brain region and a peripheral blood vessel, for example to correlate central release of a hormone with blood levels (e.g. Wotjak et al., 1996).
5. Measurement of prostaglandin levels in brain following induction of fever (e.g. Sehic & Blatteis, 1996), and in spinal cord following a noxious stimulus (e.g. Malmberg & Yaksh, 1995).
6. Pharmacokinetic measurements such as the analysis of the concentration of a drug in the brain following systemic administration (e.g. Pan & Lee, 1995).

Obviously, this is not an exhaustive list but serves to demonstrate the range in applicability of *in vivo* microdialysis.

PITFALLS OF THE TECHNIQUE

It should be clear that *in vivo* microdialysis has a wide variety of uses, both in terms of measuring neurotransmitter release and also as a drug delivery system/pharmacokinetic tool. This is all possible due to the chemical specificity of the technique; however, this also becomes a potential pitfall of the technique. As has been previously mentioned, the chemical specificity derives from the sample analysis, therefore, one needs to maintain a specific and sensitive means of measuring the desired analyte, which is not always straightforward.

In addition to requiring a reliable sample analysis system, *in vivo* microdialysis also has two other major shortcomings. Firstly, anatomical resolution, although greater than push-pull cannulae, is limited by the external diameter of the membrane (~250–300 μm). This means that specific sampling from particularly small brain nuclei is difficult. Secondly and more importantly, *in vivo* microdialysis suffers from poor temporal resolution. Perfusion of aCSF through the probe is typically limited to 1–2 μl/min which results in sample time of around 10–20 min. Such long sampling makes measurement of transient changes in release hard to resolve, and as a result prolonged stimuli or the use of transport inhibitors may be necessary. The development of more sensitive assay systems, such as microbore HPLC (Smolders et al., 1995) and capillary electrophoresis (Bergquist et al., 1996) has gone some way towards addressing the issue of reducing sampling times and thereby increasing temporal resolution. Another point to be aware of is the need for stable anaesthesia in acute preparations, as neurotransmitter release will vary with the level of anaesthesia.

RECENT PAPER

A short communication by Wotjak et al. (1996), describes an interesting experimental paradigm in which two different applications of microdialysis are used within the same animal. Anaesthetized rats (urethane) were implanted with a microdialysis probe in either the supraoptic (SON) or paraventricular nucleus (PVN) of the hypothalamus and also another dialysis probe was placed into the jugular vein. In addition, a push-pull cannula was inserted into the median eminence. Two basic strategies were employed: either perfusion of a depolarising stimulus of high potassium or perfusion of hypertonic saline. In both cases the release of arginine vasopressin (AVP) was measured in the blood, median eminence and either the SON or PVN. When a high potassium stimulus was applied to the PVN, an increase in AVP was only apparent in the PVN itself; however, when a high potassium stimulus was applied to the SON, increases in AVP release were measured in the SON, median eminence and also in the blood. Similarly, hypertonic saline perfusion in the SON resulted in increased levels of AVP in the SON and in the blood. These findings were suggested to indicate a role for AVP in the control of both the hypothalamic-neurohypophysial system and the hypothalamic-pituitary-adrenal axis; however, further experiments would be required to fully address this hypothesis.

REFERENCES AND FURTHER READING

Benveniste, H. and Hettemeier, P. C. (1990) Microdialysis: theory and application. *Prog. Neurobiol.* **35**, 195–215.

Bergquist, J., Vona, M. J., Stiller, C. O., O'Connor, W. T., Falkenberg, T. and Ekman, R. (1996) Capillary electrophoresis with laser-induced fluorescence detection: a sensitive method for monitroing extracellular concentrations of amino acids in the periaqueductal grey matter. *J. Neurosci. Meth.* **65**, 33–42.

Chen, S. Y., Burger, R. L. and Reith, M. E. A. (1996) Extracellular dopamine in the rat ventral tegmental area and nucleus accumbens following ventral tegmental infusion of cocaine. *Brain Res.* **729**, 294–296.

Fedele, E. and Raiteri, M. (1996) Desensitization of AMPA receptors and AMPA-NMDA receptor interaction: an *in vivo* cyclic GMP microdialysis study in rat cerebellum. *Br. J. Pharmacol.* **117**, 1133–1138.

Hsiao, J. K., Ball, B. A., Morrison, P. F., Mefford, I. N. and Bungay, P. M. (1990) Effects of different semipermeable membraneson *in vitro* and *in vivo* performance of microdialysis probes. *J. Neurochem.* **54**, 1449–1452.

Justice Jr, J. B. (1993) Quantitative microdialysis of neurotransmitters. *J. Neurosci. Meth.* **48**, 263–276.

Lawrence, A. J. and Jarrott, B. (1994) L-Glutamate as a neurotransmitter at baroreceptor afferents: evidence from *in vivo* microdialysis. *Neuroscience* **58**, 585–591.

Malmberg, A. B. and Yaksh, T. L. (1995) Cyclo-oxygenase inhibition and the spinal release of prostaglandin E2 and amino acids evoked by paw formalin injection: a microdialysis study in unanaesthetized rats. *J. Neurosci.* **15**, 2768–2776.

Pan, W. H. and Lee, J. C. (1995) Determination of microdialysis extraction fraction of cocaine by the no net flux method under high, low and zero steady-state concentrations in rats. *Brain Res.* **686**, 249–253.

Robinson, T. E. and Justice Jr, J. B. (1991) *Microdialysis in the Neurosciences*. Amsterdam: Elsevier.

Sehic, E. and Blatteis, C. M. (1996) Blockade of lipopolysaccharide-induced fever by subdiaphragmatic vagotomy in guinea-pigs. *Brain Res.* **726**, 160–166.

Smolders, I., Sarre, S., Michotte, Y. and Ebinger, G. (1995) The analysis of excitatory, inhibitory and other amino acids in rat brain microdialysates using microbore liquid chromatography. *J. Neurosci. Meth.* **57**, 47–53.

Wotjak, C. T., Kubota, M., Kohl, G. and Landgraf, R. (1996) Release of vasopressin from supraoptic neurons within the median eminence *in vivo*. A combined microdialysis and push-pull perfusion study in the rat. *Brain Res.* **726**, 237–241.

CHAPTER 17

ELECTRORETINOGRAPHY

A. Leon
Division of Neuroscience, John Curtin School of Medical Research, Australian National University

INTRODUCTION

Electroretinography is a recording technique which is used to assess the electrophysiological function of the retina. It has clinical applications in the diagnosis of certain ocular diseases, and it is also used as a research tool. In its simplest form the recording, an electroretinogram (ERG), represents the massed electrical potential generated by the retina in response to a flash of light.

WAVEFORM OF THE ERG

The components of a typical ERG, recorded in response to a high intensity flash of light, are illustrated in Figure 17.1. Assessment of the electroretinographic recording involves measurement of parameters such as the amplitude, time-to-peak, threshold, and latency of the different components of the ERG waveform. The ERG comprises a short latency cornea-negative a-wave (the amplitude of which is measured from the baseline to the trough of the a-wave) followed by a larger cornea-positive b-wave (the amplitude of which is measured from the trough of the a-wave to the peak of the b-wave). In DC recordings a much slower cornea-positive c-wave is also found. Superimposed on the b-wave (particularly on its ascending limb) are several small wavelets known as oscillatory potentials. The composite flash ERG represents a summation of waveforms generated by different processes critically dependent on the initial photoreceptor response. The initial electrical event in the ERG, and one which is not shown in Figure 17.1, is a very rapid, low amplitude component which precedes the a-wave and occurs within the first few milliseconds—the early receptor potential (ERP). The ERP represents photoisomerisation of rhodopsin molecules within the photoreceptor outer segments. The a-wave represents the early part of the photoreceptor response (the descending limb of the a-wave corresponds to increasing hyperpolarization of the plasma membrane of the photoreceptor outer segments as a result of closure of sodium channels). The b-wave reflects potassium ion fluxes in the extracellular space between the Müller cells and the depolarizing bipolar cells. Where present, the c-wave is mostly generated by the retinal pigment epithelium. The oscillatory potentials are thought to be generated within the inner layers of the retina.

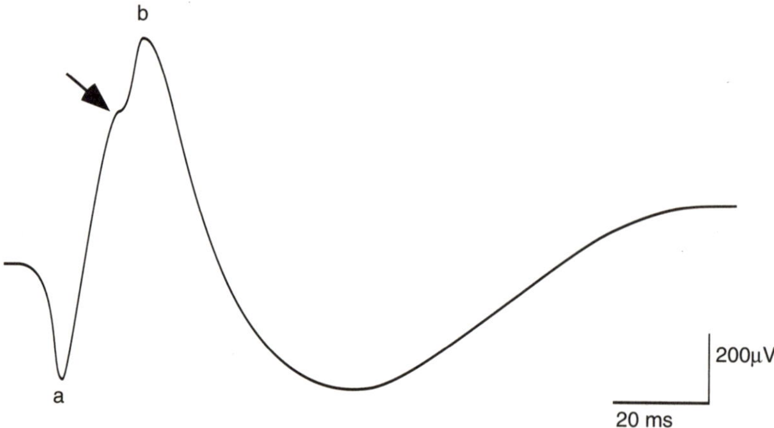

Figure 17.1 Waveform of a typical ERG (cat) recorded under scotopic conditions in response to a 10 μs high intensity flash of white light. The a-wave and b-wave are shown and the arrow indicates an oscillatory potential.

EQUIPMENT AND METHODS

Electroretinography of human subjects is usually performed on conscious individuals with the aid of a head-rest, whereas in unco-operative animal subjects anaesthesia is required. Some anaesthetic agents, such as halothane, depress the ERG and should be avoided. Urethane is reputedly the best agent from this point of view although there are hazards associated with its use. Ketamine and alphaxalone/alphadolone are more commonly used and both give acceptable results. In animals the anaesthesia optimally is accompanied by neuromuscular blockade which abolishes eye movement and prevents the recording of intrusive muscle potentials. The subject's pupil is dilated before electroretinography using topical applications of drugs such as tropicamide, cyclopentolate and phenylephrine. This facilitates stimulation of the entire retina. The subject is then dark-adapted for a period of 10–30 minutes so that initial recordings of rod photoreceptor function can be made (rods function best in dim light or scotopic conditions).

The ERG is usually recorded from the surface of the cornea. The recording electrode used can be a corneal contact lens type (examples are the Burian-Allen electrode and the ERG-jet disposable electrode) or one which is inserted into the lower conjunctival sac/looped over the lower eyelid (such as a gold leaf electrode). The recording electrode is often wetted with a conducting solution such as saline, sometimes combined with a methylcellulose solution. When fitting a contact lens electrode it is important to ensure that no air bubbles become trapped between the lens and the cornea since this will result in a noisy recording. The cornea is anaesthetized topically before application of the ERG electrode to conscious individuals. Fitting the electrode is performed in darkened conditions, aided by dim red light (long wavelength light provides minimal stimulation of the dark-adapting rod photoreceptors). A reference/indifferent electrode and a ground electrode are also fitted at this time. In humans these are cutaneous and in animals they tend to be subcutaneous needle electrodes. The reference electrode is most frequently positioned close to the lateral canthus (corner) of the eye or on the forehead, and the ground electrode on an earlobe or at the back of the head or neck. The ERG signal is amplified by a pre-amplifier (the band-pass settings used differ from lab to lab and depend on what aspect of the ERG is being investigated) and waveforms are displayed on an oscilloscope where they may be photographed or stored on computer disk.

ERGs are elicited through light stimuli which may be provided from a variety of photostimulators. A typical photostimulator is the Grass PS-22 which delivers brief flashes of white light from a xenon discharge tube. This photostimulator allows the investigator to vary both the flash intensity and frequency. Greater variations of flash intensity are achieved by interposing neutral density filters in the light path. Chromatic filters may be used to vary the stimulus wavelength. In order to record full-field ERGs (elicited by homogeneous stimulation of the entire retina) the flash stimuli must be 'ganzfeld'. Ganzfeld stimulation involves scattering the light before it enters the eye. This is best achieved by using a ganzfeld diffusing sphere or dome with a white interior which is illuminated by the photostimulator. The subject's eye is positioned close to a small opening in the diffusing sphere. The sphere can range in size from a ping-pong ball to 50 cm in diameter or even larger. Some investigators attempt to achieve ganzfeld stimulation by scattering the light using opaque ERG contact lens electrodes or by placing an opaque screen immediately in front of the eye. However, these techniques result in a considerable reduction in intensity of the light stimulus. Calibration of the light stimulus intensity and surface luminance within the ganzfeld dome should be performed at some stage using a photometer, although unfortunately there is no standard unit used to describe this in the electroretinographic literature. Units reported are disparate and include foot-lamberts, scotopic trolands, lumens/m^2 and candelas/m^2.

RECORDING PROTOCOLS

There are many different protocols used in electroretinography, the choice of which will depend on the precise nature of the investigation being performed. Once the subject is placed in a darkened room some investigators record single flash ERGs at regular intervals to assess the time-course of dark adaptation and to monitor changes in ERG waveform parameters during this process. ERGs are frequently first recorded in scotopic conditions (dim light) in the dark-adapted eye in response to single flashes of white light, beginning at low intensities and progressively increasing the flash intensity in 0.1 or 0.3 log unit steps. The scotopic ERG represents largely a function of the rod photoreceptors. Very low amplitude signals, or those plagued by noise, may be improved by averaging several responses, allowing sufficient time between flashes so as to maintain the dark-adapted state of the eye. Rod responses may also be recorded using short wavelength (blue) light flashes, whereas cone responses are more readily recorded using long wavelength (red) stimuli—the wavelength of the stimulus is selected as being appropriate for the different photoreceptor photopigments.

The cone photoreceptors also function best at higher light intensities. Cone ERGs may be recorded using flashes of light superimposed on a background light which has the effect of light-adapting the eye and diminishing or abolishing the rod response. Cone responses are frequently recorded using a flickering light stimulus. Whilst rods can respond only to slow rates of flicker (<20 Hz in humans), the cones are able to respond to much higher frequencies. Clinically, the cone response is often recorded at a set flicker frequency of 30 Hz. Rod and cone signals can also be separated by recording the critical flicker fusion frequency (the frequency at which the ERG signal becomes buried within the noise) at progressively increasing light intensities. In this way one derives a flicker fusion-intensity response curve that provides information on one aspect of photoreceptor temporal function. The rods and cones yield separate branches of this curve. There are many other, often complex, electroretinographic protocols used in research and a description of these is beyond the scope of this chapter; the interested reader is referred to more specialized texts.

ANALYSIS OF ERG RECORDINGS

There are many ways of analyzing electroretinographic data. Changes in the ERG waveform can be obvious. Certain components may be absent or reduced, for example, in congenital stationary night blindness the scotopic ERG has a negative waveform with a normal a-wave, but a very diminished b-wave, probably due to a block of scotopic transmission to postreceptoral structures at the level of the rod synapse. Changes in the ERP may be observed in diseases characterized by abnormalities of the rhodopsin molecule. In other retinal diseases the ERG waveform may be essentially normal, but there may be reductions in response amplitudes and prolonged times-to-peak (e.g. in some types of retinitis pigmentosa). Plotting ERG amplitudes and times-to-peak against light stimulus intensity yields useful information on photoreceptor function. Shifts in the intensity-amplitude response curve towards higher light intensities are expected in diseases involving a generalized loss of photoreceptors or rhodopsin (reduced quantal catch). Alterations in the slope of both the intensity-amplitude response function and the intensity-temporal response function may be seen in diseases involving abnormal phototransduction mechanisms. Shifts in the rod and cone branches of the flicker fusion response curve may occur in some retinal diseases and there may even be absence of part of the curve if a whole population of photoreceptors has been lost (e.g. in rod or cone dystrophies).

RECENT PAPER

Breton et al. (1994) established a direct relationship between the ERG and biochemical events within rod photoreceptors. They recorded ERGs from several dark-adapted human subjects including normals and patients affected by different retinal dystrophies (e.g. retinitis pigmentosa), and analyzed these responses with a model of the activation steps of the G-protein cascade. The authors concluded that the rod photocurrent is directly reflected by the a-wave of the ERG, and the rising phase kinetics of this wave are accurately represented by their simple model of the G-protein cascade. This study allows deductions to be made about the pathology of phototransduction in patients affected with retinal diseases.

REFERENCES AND FURTHER READING

Armington, J. C. (1974) *The Electroretinogram*. New York: Academic Press.

Berson, E. L. (1981) Electrical phenomena in the retina. In *Adler's Physiology of the Eye* (7th edn), edited by R. A. Moses, pp. 466–529. St Louis: C. V. Mosby Company.

Birch, D. G., Hood, D. C., Nusinowitz, S. and Pepperberg, D. R. (1995) Abnormal activation and inactivation mechanisms of rod transduction in patients with autosomal dominant retinitis pigmentosa and the pro-23-his mutation. *Invest. Ophthalmol. Vis. Sci.* **36**, 1603–1614.

Breton, M. E., Schueller, A. W., Lamb, T. D. and Pugh, E. N. (1994) Analysis of ERG a-wave amplification and kinetics in terms of the G-protein cascade of phototransduction. *Invest. Ophthalmol. Vis. Sci.* **35**, 295–309.

Brunette, J. -R. (1973) A standardizable method for separating rod and cone responses in clinical electroretinography. *Am. J. Ophthalmol.* **75**, 833–845.

Gouras, P. (1970) Electroretinography: some basic principles. *Invest. Ophthalmol.* **9**, 557–569.

Marmor, M. F., Arden, G. B., Nilsson, S. E. G. and Zrenner, E. (1989) Standard for clinical electroretinography. *Arch. Ophthalmol.* **107**, 816–819.

SECTION 7

HISTOCHEMICAL TECHNIQUES

CHAPTER 18

ANATOMICAL TRACING OF NEURONAL CONNECTIONS

L. R. Marotte
Developmental Neurobiology, Research School of Biological Sciences, Australian National University

INTRODUCTION

There are now a large number of compounds in use for tracing neuronal connections. The vast majority depend on uptake and active transport by the cell. Depending on the compound, they may be taken up by the cell body and transported anterogradely to the axonal terminals (e.g. *Phaseolus vulgaris*), or taken up by the terminals and transported retrogradely to the cell body (e.g. fluorescent dyes such as Fast Blue and diamidino yellow), or transported in both directions (e.g. horseradish peroxidase (HRP) and carbocyanine dyes such as 1,1′-dioctadecyl-3,3,3′,3′ tetramethylindocarbocyanine perchlorate, (DiI)). In the latter situation an injection into a region of the CNS will reveal projections both to and from the injection site.

Since the passage of these substances depends on uptake and active transport by the cell they must be applied in the living animal. After a survival period, generally of the order of 24–48 hours, to allow transport from the injection site to the terminals and/or cell bodies, the animal is deeply anaesthetized and perfused briefly through the vascular system with physiological saline to remove the blood and then with a fixative to preserve the tissue.

To visualize the label, sections of the tissue have to be cut into thin layers and viewed in a microscope. However, in special cases it is possible to use whole mounts of tissue when it is relatively thin and can be flattened to be examined under a microscope. For example, whole mounts of the retina can be made after removing the sclera and pigment epithelium and making radial cuts to flatten it onto a microscope slide. This is a particularly convenient way to examine the classes of retinal ganglion cells that project to different central areas. Ganglion cells can be retrogradely labelled from injections in different visual nuclei in different animals and the projection patterns of retinal ganglion cell classes analyzed.

After sectioning or preparing a whole mount, tissue containing fluorescent dyes can then be visualized without further procedures, using a fluorescence microscope and filters of appropriate wavelength for the dye used. When such dyes fluoresce at wavelengths that are sufficiently different from each other, more than one dye can be injected in the same animal at, say, two different injection sites. Cells in a particular region which project to these two areas can then be examined using the appropriate filters to see whether they are labelled with only one dye, indicating separate populations project to the two areas. If cells contain both dyes this indicates that they send a branch to both areas. Unfortunately these dyes fade on viewing so analysis must be carried out immediately.

Non-fluorescent tracers such as HRP require further procedures to be visualized in the tissue. In the case of HRP this is a histochemical method dependent on the enzyme activity of HRP to produce a visible reaction-product. Thus, there is amplification of the signal making it a more sensitive method than when there is just a 1:1 visualization of the transported tracer. This enables sparse connections to be demonstrated more easily. In contrast to the fluorescent dyes HRP produces a permanent reaction-product.

In the last few years the advent of fluorescent carbocyanine dyes has made it possible to trace connections in fixed tissue. As well as being actively transported *in vivo*, these dyes are lipophilic and in fixed tissue dissolve and travel along the cell membrane. This has made tracing neuronal connections in prenatal fixed tissue a relatively simple procedure whereas, in the past, it was either extraordinarily difficult or impossible to do *in vivo*. Procedures for *in vivo* tracing using HRP and *in vivo* and *in vitro* tracing using carbocyanine dyes are briefly described below.

HRP (MESULAM, 1982)

Application

HRP is supplied as a powder. It is generally injected as a 30% solution in distilled water or physiological saline, using a microsyringe and volumes in the order of 0.01–0.1 μl. Smaller and more concentrated injection sites in the central nervous system are obtained if it is linked to the plant lectin, wheatgerm agglutinin (WGA-HRP). This lectin binds to the cell surface, both preventing the spread of HRP and increasing its uptake by the cell. Discrete injection sites confined to the nucleus of interest are necessary. Spread to adjacent structures confuses interpretation of the results. HRP can also be taken up by damaged axons passing through the injected area so this should be kept in mind when interpreting the results.

Extremely small deposits in developing animals can be obtained by drying a concentrated solution of HRP onto the tip of a microelectrode which is then inserted into the tissue to dissolve the HRP. Alternatively a solution can be drawn into a micropipette by capillary action and inserted as above. Tip diameter can be varied to alter the size of the deposit.

Histochemistry

After 1–2 days survival the animal is deeply anaesthetized and perfused, initially with buffer or saline, to remove the blood. This is important as the endogenous peroxidase activity of red blood cells will also result in the production of reaction-product which can obscure the reaction-product produced by the transported HRP. Fixation is brief so that enzyme activity is not compromised. The standard histochemical method for light microscopy is the tetramethylbenzidine (TMB) method (Mesulam, 1982). This is very sensitive and, as long as the method is followed as Mesulam describes, it is extremely reliable. Unfortunately the reaction- product produced is soluble in alcohol so it is not suitable for situations that require prolonged dehydration such as preparation of tissue for electron microscopy. Other methods are available which are suitable for electron microscopy (Mesulam, 1982).

Label

With the TMB method, anterograde label in the light microscope appears as fine brown dust. Retrograde label is in the form of granules confined to the cell body and the base of dendrites. Often axons are not well labelled and, when present, label is granular. The TMB method is not suitable for revealing the dendritic trees of retrogradely labelled cells.

CARBOCYANINE DYES (GODEMENT ET AL., 1987)

A number of these dyes are now available (DiI, DiA, DiQ) which fluoresce at different wavelengths meaning several dyes can be used in the one experiment. They are transported both *in vivo* and *in vitro*.

In vivo

Dyes are insoluble in water but can be injected as a 5–10% solution in dimethylformamide using a microsyringe. Depending on the accessibility of the site, an undissolved crystal may be inserted into the tissue. Because they are insoluble in water, there are less problems with spread of label at the deposit site than with HRP. The dyes are transported anterogradely and retrogradely at a similar rate to HRP. After survival periods of up to several days, the animals are deeply anaesthetized and perfused. Fixative should not contain glutaraldehyde as this autofluoresces, increasing the background. If tissue is to be sectioned a vibratome, rather than a freezing microtome, should be used as frozen sections also increase the background. Unlike the granular and unreliable filling of axons by HRP, the dyes fill the axon along its entire course with solid label. Thus from a deposit in the retina the axon can be followed across the retina to the optic disc, along the optic nerve and tract and across the superior colliculus to its termination site. As with other fluorescent dyes, there is fading on viewing, so analysis must be done immediately.

In vitro

In fixed tissue small undissolved crystals of the dye are generally inserted into the tissue. Because the dyes travel along any membrane they come in contact with, it is important that no stray crystals escape from the site. During transport tissue is kept in the fixative, in the dark at room temperature. Disadvantages are that transport times range from weeks to months depending on the distance of the pathway being traced. Transport at 37°C increases the rate of transport but may be detrimental to the tissue. It has also been claimed that the dyes do not transport well *in vitro* in myelinated axons, making them unsuitable for most mature nervous system pathways. The distances involved in adult tissue and consequent lengthy transport times also make them a less attractive choice in this situation.

The entire axon and dendritic tree of cells are labelled giving a Golgi-like picture of the pathway. There can be advantages and disadvantages in this. The dendrites may obscure the details of the afferent input to the area. Another serious drawback which must be taken into account in interpreting results is that dye can be transported transcellularly, either when there are tight junctions or perhaps just close apposition of membranes. Tight junctions are known to exist in the developing nervous system. If a cell body is labelled in a region where there are labelled axon terminals from the deposit site, it should not be assumed that this cell necessarily projects to the deposit site as transcellular labelling from dye in the axon terminals to cell bodies in the region may have occurred. A permanent reaction product can be produced by photoconversion of the dye to a stable reaction product. This is only feasible for very small areas of tissue as the tissue must be fluoresced using a high power objective. The method can also be capricious.

RECENT PAPER

A recent paper (Clascá et al., 1995) which has used the fluorescent dyes, DiI in fixed tissue and Fast Blue *in vivo*, to examine development of the cerebral cortex, highlights the advantages and disadvantages of the different tracing methods. There has been controversy over which cells first leave the cortex and enter the thalamus during development. Results differed depending on the tracer used. This paper shows that previous work, using DiI in fixed tissue, most likely misidentified these cells due to transcellular labelling of the dye from labelled afferents in the cortex. The use of both dyes in this study shows that the few cells retrogradely labelled from the thalamus with DiI early in development are not labelled after similar injections of Fast Blue, a dye which does not appear to undergo transcellular transport. It is not until later in development that cells genuinely projecting to the thalamus can be detected using Fast Blue.

REFERENCES AND FURTHER READING

Clascá, F., Angelucci, A. and Sur, M. (1995) Layer-specific programs of development in neocortical projection neurons. *Proc. Natl. Acad. Sci.* **92**, 11145–11149.

Godement, P., Vanselow, J., Thanos, S. and Bonhoeffer, F. (1987) A study in developing visual systems with a new method of staining neurons and their processes in fixed tissue. *Devel.* **101**, 697–713.

Mesulam, M. -M. (1982) *Tracing Neuronal Connections*. New York: John Wiley and Sons.

CHAPTER 19

IMMUNOHISTOCHEMISTRY

P. Cooper
Division of Botany and Zoology, The Faculties, Australian National University

INTRODUCTION

Antibodies can recognize either proteins, peptides, or amines involved in either neurotransmission or neuroreception. By using antibodies (either commercially available or developed by yourself or an associate), it is possible to localize the presence of neurotransmitters or receptors in any tissue of interest by using the indirect method of localisation (Figure 19.1A). The procedure requires: (1) incubating the tissue with the antibody (Ab) so that it binds to antigen present in the tissue; (2) washing the tissue so that only bound antibody remains; and (3) exposing the tissue with the antibody to a second antibody which recognizes the first antibody, and is conjugated to either a fluorescent chemical (e.g. fluorescein isothiocyanate (FITC)) or enzyme system (e.g. horseradish peroxidase (HRP)) which produces a visible reaction product. The tissue may then be examined under fluorescent or light microscope. The secret of the technique is to maximize the signal (positive staining) while minimising the background staining (false positive staining). This technique has become extremely important for determining the three-dimensional location of a wide variety of neurons within any nerve tract.

An alternative procedure to the above simple indirect method is the so-called 'sandwich' method, which incorporates a third antibody step (Figure 19.1B). In this procedure following the primary antibody application, a second antibody which recognizes the first antibody is applied. This is followed by a third antibody raised in the same animal as the primary antibody, but conjugated to peroxidase-antiperoxidase system (Sternberger, 1979). The response is more sensitive than the simpler system of a second antibody conjugated to HRP, but the reason for this is not clear. Other procedures involve avidin-biotin complexes which also improve sensitivity.

In each procedure, it is important that antibodies must be selected for their specific recognition characteristics. For example, if the primary antibody is raised in rabbit, then the second antibody must be raised against rabbit serum in some other animal, typically sheep or goat, although rodents are also popular laboratory antibody 'factories'. This is especially critical in the current trend to multiple labelling regimes, as two or more primary antibodies are used to look for cells with co-localized chemicals. In these procedures, the primary antibodies should be raised in different animals (e.g. rabbits and rats in the case of dual labelling), with the secondary antibodies having different materials conjugated (e.g. anti-rabbit IgG conjugated with fluorescein and anti-rat IgG conjugated with HRP) selected to ensure that desired neurochemicals are distinguished.

A. INDIRECT METHOD

Conjugate (either fluorescent or enzyme marker)

2° Ab

1° Ab

Tissue Antigens

B. PAP METHOD

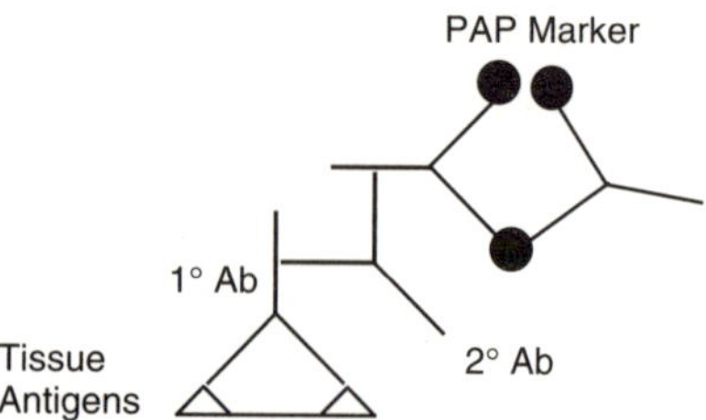

Figure 19.1 A: Representation of the binding of the primary antibody (1° Ab) to cellular antigens and the subsequent binding of a labelled secondary antibody (2° Ab) to the primary antibody. The label on the secondary antibody can be either fluorescent or enzymatic (e.g. horseradish peroxidase or alkaline phosphatase). B: Representation of the three-step procedure for the peroxidase-antiperoxidase (PAP) method. The primary antibody recognizes the cellular antigens, then a secondary antibody recognizes the primary antibody and acts as a bridge to the tertiary antibody which has the PAP reaction site. The primary and tertiary antibodies should be raised in the same animal (i.e. rabbit) and the secondary antibody should be raised against the immunoglobin G (IgG) of the animal from which the primary antibody was raised. In this case, anti-rabbit IgG should be raised in either rodent, sheep or goat (redrawn from Kuhlmann, 1984).

PROCEDURE

The procedure is extremely simple and commercial kits are often available for many chemicals of interest to neuroscientists. The tissue of interest can be treated as a whole unit if the tissue is small enough or larger tissues can be sectioned, either in thin (<15 μm) or thick (>50 μm) sections. The tissue must first be fixed, typically in freshly made 4% paraformaldehyde or Bouin's fixative. We dissect the tissue in a saline solution, then transfer the tissue to a watch glass containing the fixative for a specific length of time. Following fixation, the tissue can either be used for the whole-mount procedure or embedded in paraffin or frozen, although embedding in gelatine has also been successful. When freezing, a cryoprotectant must be used. Typically fixation causes some morphological changes, especially shrinkage of tissue. Paraffin embedded material is sectioned using a conventional microtome, section thickness being 5–10 μm; frozen material is sectioned using a cryostat. For the thickest sections, gelatine-embedded material is cut with the aid of a vibratome. In all cases, sectioned material is transferred to slides which have been coated either with gelatine or poly-L-lysine (1–2% solution) to ensure tissues adhere to slides during subsequent treatment. Following coating, the slides must be allowed to dry before transferring the tissue sections.

Whole mounts

Following fixation, whole mounts must be made permeable to the antibody solutions. The tissues are first transferred to small vials. The paraformaldehyde must be removed from the tissue by washing 2–3 times with phosphate-buffered saline (PBS), then the tissue is made permeable by treatment with methanol. We do this by treating for 5 min, 70% MeOH (in PBS), then for 60 min in 100% methanol (MeOH), then for 5 min in 70% MeOH (in PBS), and then washing with PBS 2–3 times. We then wash the tissues twice in PBT (PBS with 0.2% bovine serum albumin and 0.1% Triton X-100), which acts both to make membranes permeable, but also is an initial step in blocking non-specific binding of the antibody. This is followed by incubation for 30 min in PBT with normal goat serum (NGS). During this step, we place the tissues on a shaker to ensure complete mixing of the tissue and material. The NGS is necessary for further blocking of non-specific binding. At this point, the primary antibody (which is produced against the chemical of interest) is diluted in PBT+NGS, and added to the tissue. The dilution varies for various antibodies and is detailed in kits. Alternatively, if the antibody is obtained non-commercially, some experimentation may be necessary to determine the appropriate dilution. This may require running a series of dilutions on the first attempt to see which is most appropriate. Normally, the series would be 1:5000, 1:2000, 1:1000, 1:500, although if the concentration of the antibody is known this can be better reported as a dilution in terms of mg Ab per ml solution. The secret with the dilution is to use the lowest dilution possible and still obtain a reaction. As the concentration of antibody increases, more non-specific binding can occur and false positives can appear.

The primary antibody is allowed to incubate with the tissue from overnight to 24 h at 4°C. Following this incubation, the primary Ab is removed by washing the tissue in PBT (3 times), and then washed for longer periods (45 min–1 h) on a shaker (2 times). A second incubation in PBT+NGS for 30 min follows, after which the secondary Ab (the one which is raised to the serum of the animal in which the first Ab was raised) is diluted in PBT+NGS (again this varies, but we use 1:40), and the tissue is incubated in this solution overnight. Higher concentrations of the secondary Ab may reduce the time for incubation. The tissue is then washed 3 times over 15 min in PBT, and then 4 times for 30 min each in PBT on a system to ensure complete inversion. At this point the technique diverges if the secondary Ab is conjugated to HRP or a fluorescent chemical (e.g. FITC, tetramethylrhodamine isothiocyanate (TRITC), Texas Red, rhodamine).

For fluorescent chemicals, the tissue is rinsed 3 times over 15 min in PBS, then cleared in 70% glycerol (70% glycerol in PBS) overnight. The tissue is then mounted in 0.1% p-phenylenediamine (antifade agent) in glycerol. We mount the tissue on a conventional slide under a cover slip which is supported by dried clear nail polish on the four corners. After orientating the tissue, the cover slip is placed onto the supporting corners and sealed in place with nail polish.

For HRP conjugated Ab, after washing with PBT, 200 μl of PBT is added with 100 μl diaminobenzidine (DAB) (0.5–1 mg/ml) and 8% nickel chloride (4 μl per 0.5 ml DAB solution), which intensifies the reaction, forming a darker color. Incubate for about 10 min, then add 3 μl 2% hydrogen peroxide. Allow color formation to proceed for 1–5 min (or less if high background); this step requires observation of the reaction under a microscope, as background can increase very quickly. Rinse with PBT several times to stop reaction. Rinse the tissue at least 3 times over 15 min in PBS, then clear the tissue as above or with methyl salicylate and mount.

Tissue sections

Following cryostat sectioning the slides can be immediately processed or can be stored in a freezer over silica gel. The latter procedure is actually better as the tissue adherence to the slide may improve as a result. The slides are brought to room temperature slowly over 7–8 minutes, then incubated in 0.5% Triton X-100 in PBS overnight. Wash the slides in normal PBS 2–3 times over 10 min, then apply 50 μl NGS to each tissue section. Incubate the slides for 30 min at room temperature in closed chamber at 100% relative humidity (moist paper towel on the bottom of a desiccator works well). Wash 3 times over 10 min with 0.1 M PBS. Blot the slides, but ensure that the tissue is not touched nor allowed to dry out. Apply the primary Ab (dilution depends upon concentration of undiluted Ab, dilution should be in PBS+NGS). Incubate overnight in closed humid chamber at 4°C. Wash slides 3 times with 0.1 M PBS over 10 min. Blot slides, then apply secondary antibody as before (again dilution depends upon concentration of undiluted secondary Ab). Incubate tissues again in closed humid chamber for 60 min-overnight, depending upon Ab dilution. Wash slides 3 times in PBS, blot and allow to air dry. Follow the same procedure as above for HRP reaction. Sections can be counterstained with haematoxylin and differentiated with acid alcohol, then dehydrated through a series of ethanols (50–100% EtOH). Clear in glycerine or methyl salicylate or other mounting medium. Use antifade if using fluorescently-labelled secondary Ab, although permanent slides can be made if either HRP- or alkaline phosphatase-conjugated secondary Ab is used.

The procedure for wax embedded tissues is similar, but the tissues must be de-waxed and rehydrated before proceeding with staining protocol.

Controls

The best procedure to follow in controls is to use primary Ab which has been incubated with the chemical of interest overnight in PBS. We dilute the Ab 1:1000 in PBS, then add 30 nmoles of the material. This pre-absorbed Ab is then substituted for the untreated Ab for the controls. Other control procedures may include simply omitting the primary Ab in the first incubation step.

Using a tissue known to contain the antigen of interest as a positive control is beneficial to verify the technique, particularly if a negative result on the experimental tissue is obtained.

MICROSCOPY

The procedure adopted (sections versus whole-mount and fluorescence versus HRP) may depend upon the microscopy system available. Fluorescence requires either a fluorescent microscope or confocal scanning laser microscope (CSLM), the latter system is also useful for whole-mount viewing as optical sections can be obtained (Figure 19.2).

If either of these microscopes are not easily available, then the HRP method may be the only option. Ordinary Köhler illumination is all that is needed to examine

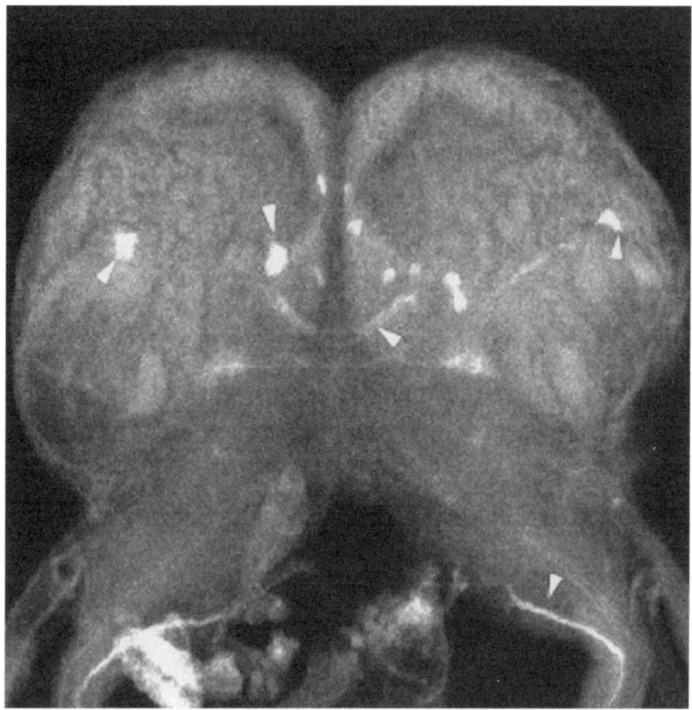

Figure 19.2 Whole mount of caterpillar (*Helicoverpa armigera*) brain. Primary antibody was rabbit anti-serotonin, secondary antibody was FITC-conjugated sheep anti-rabbit IgG. Image was taken on a Leica confocal microscope, cropped and orientated with Adobe® Photoshop, and then printed on a high-density laser printer. Both cells and axons, shown by arrows, are distinguishable in the image.

HRP-stained material. Magnification should be adjusted to ensure that the best detail is obtained. Currently, video capture systems in association with light microscopes have improved the ability to capture a series of sections in the Z-axis, which can be compressed using computer software to yield a final composite image which is equivalent to the high resolution of the confocal microscope optical section technique.

POSSIBLE PITFALLS

When purchasing commercially available antibodies, it is important to ensure that the antigen in the animal in which they were raised is the same as that in your experimental tissue. For example, many antibodies to neurohormones are raised from the mammalian antigen, but the amino acid sequence may be different from that in another vertebrate, such as a bird. If the amino acid sequence in an experimental animal is not known, there is always a possibility that it differs from that of the antigen used to raise the commercial antibody, in which case the cross-reactivity may be low so the antigen/antibody complex will not form. Alternatively, antibodies may have extensive cross-reactivity to other proteins or peptides than the ones of interest and false positives can result.

REFERENCES AND FURTHER READING

Balthazart, J., Absil, P., Foidart, A., Houbart, M., Harada, N. and Ball, G. F. (1996) Distribution of aromatase-immunoreactive cells in the forebrain of Zebra finches (*Taeniopygia guttata*): implications for the neural action of steroids and nuclear definition in the avian hypothalamus. *J. Neurobiol.* **31**, 129–148.

Kuhlman, W. D. (1984) *Immuno Enzyme Techniques in Cytochemistry*. Weinheim: Verlag Chemie.

Polak, J. M. and Van Noorden, S. (1987) *Introduction to Immunocytochemistry: Current Techniques and Problems*. Oxford: Oxford University Press.

Sternberger, L. A. (1979) *Immunocytochemistry*. New York: J. Wiley & Sons.

Yamada, M., Hayashi, S. Hozumi, I. Inuzuka, T., Tsuji, S. and Takahashi, H. (1996) Subcellular localization of growth inhibitory factor in rat brain: light and electron microscopic immunohistochemical studies. *Brain Res.* **735**, 257–264.

Yang, Q., Wang, S., Hamberger, A. and Haglid,K. G. (1996) Plasticity of granule cell-mossy fiber system following kainic acid induced seizures: an immunocytochemical study of neurofilament proteins. *Neurosci. Res.* **26**, 57–64.

CHAPTER 20

HIGH RESOLUTION AUTORADIOGRAPHY

S. Oleskevich
Visual Sciences, Research School of Biological Sciences, Australian National University

INTRODUCTION

Radiography is used to visualize the distribution pattern of radiation. Commonly, a source of radiation, such X-rays or gamma rays, passes through a specimen and is recorded onto photographic film. In autoradiography, the radiolabelled tissue is itself the source of radiation. Autoradiography can be used to visualize specific cell types and cell components such as proteins, nucleic acids, membrane receptors, and neurotransmitters. Once a cellular constituent has been tagged with a radioactive label, its location can be recorded and its density can be quantified.

Autoradiography is only one of several methods used to detect and measure radiation. Radiation can also be detected as electrical pulses, or as minute flashes of light amplified by scintillation counters. The advantage of autoradiography is that it is able to provide a topographic map of the radiation in contrast to pulse counting techniques which simply measure total specimen radiation.

The autoradiographic process begins with incubation of the tissue with radioactive material such as tritium (^{3}H), carbon-14 (^{14}C), phosphorus-32 (^{32}P), sulphur-35 (^{35}S) or iodine-125 (^{125}I). The tissue is fixed to preserve the location of the radiolabelled compounds, and then coated with a nuclear emulsion. Ionising radiation is emitted from the tissue and into the emulsion where it converts ionic silver to metallic silver at specialized sensitivity peaks. The accumulation of metallic silver at numerous peaks creates a latent image which is then developed into a true image of silver grain aggregates using standard darkroom techniques. The pattern of silver grain aggregates corresponds to a map of the radioactive compounds in the tissue.

High resolution autoradiography is a refined process that allows detection and analysis of nerve cells containing a specific neurotransmitter. The process optimizes the efficiency of the emulsion to record maximum radiation and still maintain a high degree of resolution. This is achieved by using low energy isotopes such as tritium whose emitting particles have limited travel in the emulsion yet high specific activity. The process has been refined for brain slices to visualize transmitter-specific nerve terminals and their projection into specific brain regions. The nerve terminals become labelled according to the neurotransmitter they release. The high affinity transport mechanisms, located at the nerve terminal and used normally to terminate transmitter action, take up the radioactive transmitter into the terminal. Only those nerve terminals capable of re-uptake (e.g. dopamine, noradrenaline, adrenaline, serotonin, GABA, glutamate, aspartate and choline) can be labelled by this process. At the light microscope (LM) level, the location of radiolabelled nerve terminals shows targeted brain regions while the number of terminals estimates the density of nerve terminals in these regions. Electron microscope (EM) autoradiography can suggest possible post-synaptic targets and the subcellular localisation of the radioactive labelling.

LIGHT MICROSCOPE AUTORADIOGRAPHY

Tissue preparation

During tissue preparation, the physiological mechanisms of re-uptake and axonal transport must remain viable for successful radiolabelling. The tissue is usually cut fresh into slices 200–500 μm thick in the presence of cold artificial cerebrospinal fluid. The slices are incubated with a radioactive compound. Tritium is a popular choice of radiolabel as it emits β-particles which have a high energy loss and a short travel distance. This makes them highly efficient at transferring the radiation energy to the emulsion only at specific areas.

The radiolabels are fixed into place with a chemical fixation using glutaraldehyde and osmium tetroxide to preserve the cytological and ultrastructural integrity of the tissue. Glutaraldehyde cross-links proteins through reaction with their amide groups while osmium tetroxide affects mostly ordered arrays of phospholipid molecules. The tissue is prepared for semi-thin sectioning using ethanol dehydration and resin embedding. Ethanol dehydration removes fats such as myelin and water (which will not mix with resin) and resin embedding provides a hard medium to support both semi- and ultra-thin sectioning. Semi-thin sections (3–5 μm) allow for optimal detection of tritiated compounds and good photographic resolution at the LM level (Descarries & Beaudet, 1983).

Image development

The semi-thin sections are coated with a thin layer of emulsion. Exposure times vary from one to four weeks with shorter durations for labelling nerve terminals and longer durations for axonally-transported material. The image is developed by placing the emulsion-coated sections in developer to amplify the minute silver deposits into silver grain aggregates. Figure 20.1 shows silver grain aggregates which label noradrenaline-containing nerve terminals in the rat hippocampus, following incubation and re-uptake of radioactive [^{3}H]-noradrenaline.

ELECTRON MICROSCOPE AUTORADIOGRAPHY

The tissue preparation and image development described above is also used for EM autoradiography. However, ultra-thin sections (90 nm-thick; silver/gold in color) are necessary for electron microscopy. A thinner layer of emulsion made of smaller diameter silver crystals is used to improve the resolution. The exposure period is usually 10 times that required for light microscopy. The ultra-thin sections are usually stained with lead and/or uranyl acetate for greater photographic contrast. Sometimes the sections are coated with a thin layer of carbon to encourage a uniform layering of emulsion and to improve the stability of the sections under the electron beam (Descarries & Beaudet, 1983). Figure 20.2 is an EM autoradiograph which illustrates the ultrastructural features of a silver-grain labelled [^{3}H]serotonin-containing nerve terminal.

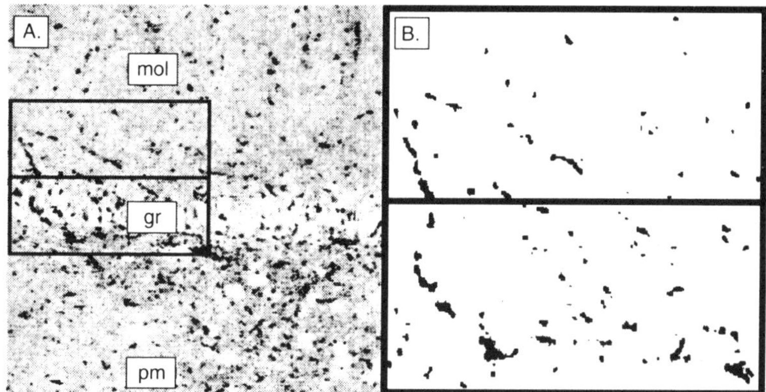

Figure 20.1 A: Light microscope autoradiograph of the distribution pattern of [^{3}H]noradrenaline-labeled nerve terminals in the dentate gyrus of rat hippocampus. B: Binary image produced by the image analysis system to estimate the density of noradrenaline terminals at 1.5×10^6 terminals per mm^3 of tissue. Magnification 160×. (Modified from Oleskevich et al., 1989.)

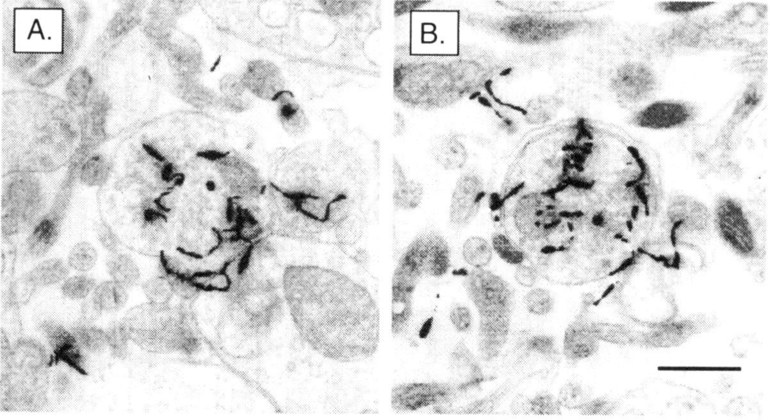

Figure 20.2 Electron microscope autoradiographs of [^{3}H]serotonin-labeled nerve terminals from Ammon's horn (A) and the dentate gyrus (B) of rat hippocampus. Both varicosities are round and exhibit small clear vesicles associated with mitochondria. Scale bar = 0.5 μm. Magnification 36000×. (Modified from Oleskevich et al., 1990.)

CONTROL EXPERIMENTS AND PRACTICAL CONSIDERATIONS

It is important to be aware of false positive results arising from alternate sources of radiation and false silver grain formation. Alternate sources of radiation include fats such as myelin which bind radioactivity non-specifically and the aldehyde groups introduced by the glutaraldehyde tissue fixative. These and other sources can be identified by exposing non-radioactive tissue under identical conditions as radiolabelled tissue and checking the number of silver grains above background. False silver grain formation may result from mechanical pressure to the emulsion. Sometimes a loss of the latent image can occur but this can be evaluated by using very radioactive tissue and checking for uniform grain density.

RECENT STUDIES

High resolution autoradiography has been used recently to assess whether certain radiolabelled substances can measure the density of serotonin nerve terminals in positron emission tomographic studies of living brain (Descarries et al., 1995). A comparison of [^{3}H]-serotonin labelled nerve terminals, quantified by high resolution autoradiography, and [^{3}H]-cyanoimipramine and [^{3}H]-citalopram binding sites, suggests that these radioligands provide a useful marker for the density of serotonin nerve terminals in living brain tissue.

REFERENCES AND FURTHER READING

Descarries, L. and Beaudet, A. (1983) The use of radioautography for investigating transmitter-specific neurons. In *Handbook of Chemical Neuroanatomy, Vol. 1: Methods in Chemical Neuroanatomy*, edited by A. Björklund and T. Hökfelt, pp. 286–364. Elsevier: Amsterdam.

Descarries, L., Soucy, J. P., Lafaille, F., Mrini, A., and Tanguay, R. (1995) Evaluation of three transporter ligands as quantitative markers of serotonin innervation in rat brain. *Synapse* **21**, 131–139.

Oleskevich, S. and Descarries, L. (1990) Quantified distribution of the serotonin innervation in adult rat hippocampus. *Neuroscience* **34**, 19–33.

Oleskevich, S., Descarries, L. and Lacaille, J. -C. (1989) Quantified distribution of the noradrenaline innervation in the hippocampus of adult rat. *J. Neurosci.* **9**, 3803–3815.

Rogers, A. W. (1979) *Techniques of Autoradiography*, (3rd edn), pp. 1–429. Amsterdam: Elsevier.

SECTION 8
BIOCHEMICAL TECHNIQUES

CHAPTER 21

MEMBRANE FRACTIONATION

P. R. Junankar
Division of Biochemistry and Molecular Biology, The Faculties, Australian National University

INTRODUCTION

In order to study the properties of membrane proteins in cell-free systems it is necessary to disrupt the cell structure and then separate the resulting intracellular organelles and membrane vesicles by differential centrifugation and density gradient centrifugation. (Broken membranes reseal into small bubble-like structures called vesicles or microsomes.) Fractions enriched in a certain type of membrane or intracellular organelle may then be solubilized in a detergent and a membrane protein purified in the presence of detergent by appropriate chromatographic methods (see Chapter 22).

METHODOLOGICAL APPROACH

Tissue and cell disruption

Tissues are carefully dissected free of fat and connective tissue, then minced or finely chopped before being homogenized using one of the techniques described in Table 21.1 (and see Figure 21.1) depending on the type and quantity of the tissue.

Protease inhibitors

It is important to prevent proteolytic degradation during tissue disruption and centrifugation by working at 4°C, adding protease inhibitors to the buffers and minimizing time spent at the crude homogenate stage. Commonly used inhibitors are described in Table 21.2.

Differential centrifugation

Intracellular organelles and membrane vesicles are separated by centrifugation. A centrifuge is a machine that spins a device called a rotor. The rotor holds tubes and bottles containing the suspended cells or fragments and spins them around a central axis. The heavier particle form 'pellets' at the bottom of the centrifuge tubes. Separation depends on the rate at which particles sediment when subjected to a centrifugal force generated by the spinning of the rotor and is a function of particle density and size as well as the

Table 21.1 Devices for tissue homogenization and cell disruption

Device	*Type of tissue*	*Capacity*	*Comments*
Waring blender	Fibrous, e.g. muscle	250 ml – 5 l	Industrial version of kitchen blender
Polytron, Ultra Turrax	Fibrous and soft tissues	0.2 ml – 2 l	Shafts with circular saw tips available in various sizes (Figure 21.1)
Potter-Elvehjem homogenizer	Softer, chopped tissues, e.g. brain, liver	5–300 ml	Pestle attached to motor (Figure 21.1)
Dounce homogenizer	Cells after lysis, resuspension of cells, organelles and vesicles after pelleting	1–100 ml	Hand-held; tight-fitting to disrupt cells, loose-fitting to resuspend organelles (Figure 21.1)
French press	Bacterial and fungal cells, fibrous tissue after initial homogenization	litres	Cells forced through fine metal mesh under pressure
Nitrogen bombs	Cells		Cell disruption by gaseous shear

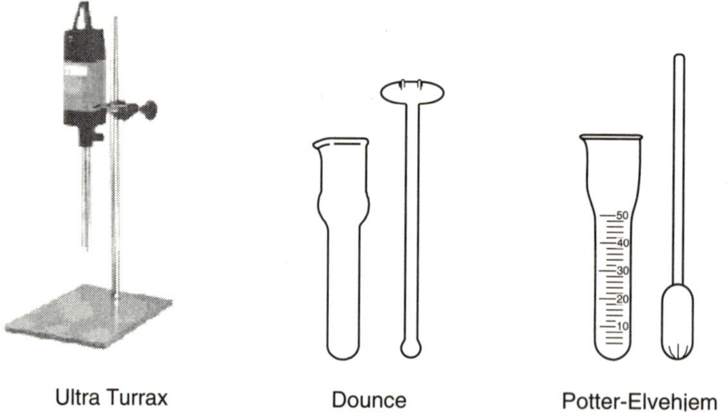

Figure 21.1 A: Ultra Turrax homogenizer. Stainless steel shafts with circular saw-like tips are rotated by the motor such that the tissue is disrupted by mechanical shear. Different sized shafts are available to process different volumes of sample. (B) Dounce homogenizer, glass vessel and pestle. Operated by hand. Different sized pestles are available for both Dounce and Potter-Elvehjem homogenizers. They can have a narrow clearance, for maximum disruption, or be more loosely fitting to preserve larger intracellular organelles. C: Potter-Elvehjem homogenizer, glass vessel and teflon pestle. Pestle is usually attached to a high-torque motor to disrupt tissue by liquid shear (but can also be operated manually). While rotating at 1500 rpm, the pestle is passed up and down the vessel 8–10 times. The vessel is usually kept chilled to dissipate the heat generated.

Table 21.2 Commonly used protease inhibitors

Inhibitor	*Specificity*	*Concentration*	*Comments*
Phenylmethylsulphonyl fluoride (PMSF)	Serine proteases	<0.5 mM	Hydrolyses in aqueous buffers (half-life ~100 min at pH 7). Prepare stock solutions in dry ethanol, dimethyl suphoxide or isopropanol
4-(2-aminoethyl)benzenesulphonyl fluoride (AESBF)	Serine proteases	0.1–1.0 mM	Stable and water soluble alternative to PMSF. More expensive
Ethylenediaminetetra- acetic acid (EDTA)	Metalloproteases	0.1-5 mM	Useful, inexpensive inhibitor
Ethylene glycol-bis(aminoethyl ether)tetraacetic acid (EGTA)	Ca^{2+}-activated proteases e.g. calpain	0.1–1 mM	
Leupeptin	Calpain, papain, plasmin, trypsin	1–100 μg/ml	
Pepstatin A	Carboxyl proteases e.g. pepsin, renin	1–10 μg/ml	Dissolve in dry ethanol or methanol
Antipain	Cathepsin B, papain, trypsin	1–10 μg/ml	
Aprotinin	Trypsin, chymotrypsin, kallikrein, plasmin	1–10 μg/ml	
Benzamidine	Serine proteases	1–10 mM	
Trypsin inhibitors (Types I–IV)	Chymotrypsin, trypsin	Depends on source	From egg white, soybean etc.

density and viscosity of the medium in which the cells have been homogenized. The relationship between the above parameters is given by Equation 1 which has been derived from Stoke's law for the diffusion of particles in a medium,

$$v = \frac{d^2(\rho_p - \rho_l)g}{18\mu} \tag{1}$$

where v = velocity of sedimentation; d = diameter of the particle; ρ_p = density of the particle; ρ_l = density of the liquid; μ = viscosity of the liquid; g = relative centrifugal field (r.c.f.).

The r.c.f. is usually expressed in multiples of the force due to gravity, and depends on the speed of the centrifuge in revolutions per min (r.p.m.) and the distance from the axis of rotation, r, in cm.

$$\text{r.c.f.} = 11.18 \times r\,(\text{r.p.m.}/1000)^2 \tag{2}$$

Intracellular organelles and membrane vesicles are separated by centrifugation. Separation depends on the rate at which particles sediment when subjected to a centrifugal force and is a function of particle weight, size, and shape as well as the density of the medium in which the cells have been homogenized. The centrifugal force is usually expressed relative to g, the force due to gravity, and depends on the speed of the centrifuge (in r.p.m.) and the radius of the rotor used. (Note that samples spun at the same speeds in rotors of different sizes will experience different r.c.f.s.) Centrifuge manuals give tables or graphical conversions of r.p.m. to r.c.f.s for each rotor.

As a rough guide, the following r.c.f.s and centrifugation times are used to sediment the following cellular organelles suspended in a buffer containing 0.25 M sucrose: 600–1000 g, 5–10 min to sediment nuclei (often called a P1 pellet); 6 000–10 000 g, 10–15 min for mitochondria, lysozymes, peroxisomes (P2 pellet) and 30 000–100 000 g, 60 min for endoplasmic reticulum (also called microsomes), ribosomes, postsynaptic vesicles (P3 pellet using high speed ultracentrifuges). Plasma membranes which form sheets may be pelleted at 15 000–30 000 g, 15 min. A standard protocol is shown in Figure 21.2. Protocols often vary, for example, the first pelleting step could be omitted and P1 contain nuclei, mitochondria etc. Then P2 would refer to the microsomal fraction. Note that all the above organelles are 'enriched', not pure and may be separated from contaminating membranes of lower density by 'washing', i.e. resuspending the pellet in fresh buffer and recentrifugating at the same speed. Separation of particles of similar size requires subsequent gradient density centrifugation.

Density gradient centrifugation

Crude preparations of membranes such as synaptosomes or microsomes can be further fractionated by density using centrifugation in media such as sucrose, Ficoll and Percoll® (Pharmacia) or Nycodenz® (Nycomed, Oslo 4, Norway). The Ficoll, Percoll and Nycomed allow separations without increasing the osmotic pressure on the organelles which occurs with increasing concentrations of sucrose. In discontinuous gradient

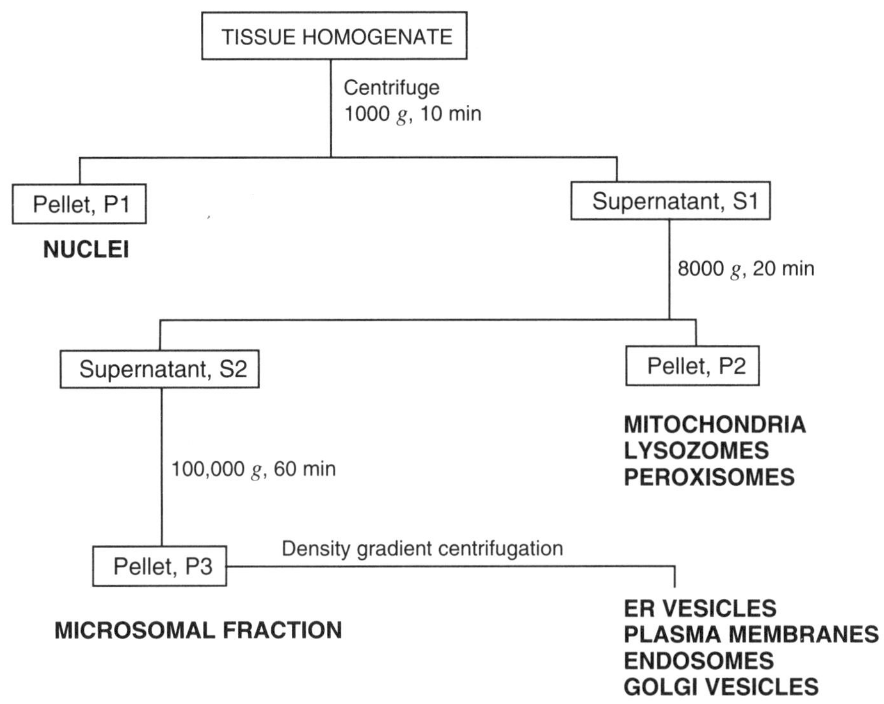

Figure 21.2 Diagram showing a standard procedure for the fractionation of subcellular organelles and membranes.

centrifugation, aliquots of the appropriate density are carefully layered one on top of the other in a centrifuge tube. For a continuous gradient, the media is mixed in a gradient maker. Samples are layered on to the top of the gradient medium. During centrifugation organelles and/or vesicles migrate to the band of the appropriate density (Figure 21.3). Linear gradients can give better separation whereas larger amounts of material can be reproducibly fractionated on discontinuous gradients. Centrifugation is usually carried out in swing-out rotors for a number of hours or overnight. Samples from a linear gradient are collected using a fraction collector; those from a discontinuous gradient are removed using a syringe or pipette. The gradient medium is then removed by centrifugation of the sample and resuspension of the pellet in the appropriate buffer.

APPLICATION TO NEUROSCIENCE

The ability to prepare synaptosomes from brain and other neural tissue has been a major impetus to the understanding of the biochemical reactions which contribute to synaptic transmission between neurons. Synaptosomes are sealed presynaptic nerve terminals which have been pinched off during homogenization of nervous tissue, sometimes with

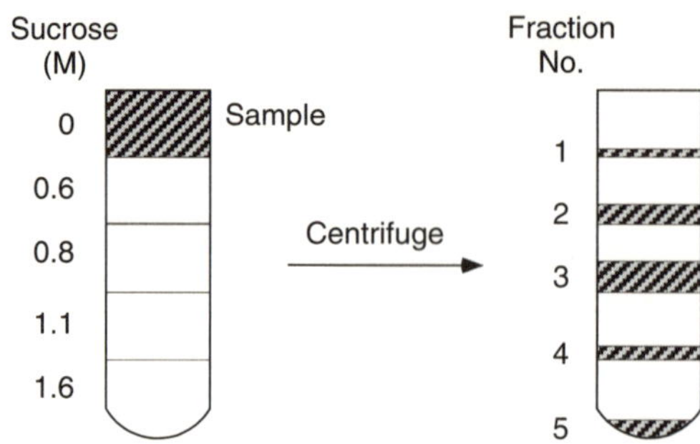

Figure 21.3 Discontinuous gradient centrifugation.

the postsynaptic membrane or even a complete dendritic spine attached. Synaptosomes can be depolarized and can release neurotransmitter in a Ca^{2+}-dependent manner. Density gradient centrifugation can be used to separate vesicles derived from presynaptic membranes from those of the postsynaptic densities. These preparations can be further disrupted in order to prepare synaptic vesicles and other synaptic components. Whittaker (1993) provides a description of the development and comparison of methods for the preparation of synaptosomes and of their use as a model for the synapse. A more recent adaptation of the traditional procedures is described in papers by Dunkley and colleagues (1988) in which they use a discontinuous Percoll gradient to purify synaptosomes from rat brain.

REFERENCES AND FURTHER READING

Fractionation of cellular organelles

Findlay, J. B. C. and Evans, W. H. (eds) (1987) *Biological Membranes. A Practical Approach*. Oxford: IRL Press.

Harris, E. L. V. and Angal, S. (eds) (1989) *Protein Purification Methods. A Practical Approach*. Oxford: IRL Press.

Centrifugation

Ford, T. C. and Graham, J. M. (1991) *An Introduction to Centrifugation*. Oxford: Bios Scientific Publishers.

Rickwood, D. (ed.) (1992) *Preparative Centrifugation*. Oxford: IRL Press.

Isolation of synaptosomes

Whittaker, V. P. (1993) Thirty years of synaptosome research. *J. Neurocytol*, **22**, 735–742.

Use of a Percoll gradient

Dunkley, P. R., Heath, J. W., Harrison, S. M., Jarvie, P. E., Glenfield, P. J. and Rostas, J. A. (1988) A rapid Percoll gradient procedure for isolation of synaptosomes directly from an S1 fraction: homogeneity and morphology of subcellular fractions. Brain Res. **441**, 59–71.

Thorne, B., Wonnacott, S. and Dunkley, P. R. (1991) Isolation of hippocampal synaptosomes on Percoll gradients: cholinergic markers and ligand binding sites. *J. Neurochem*. **56**, 479–484.

CHAPTER 22

CHROMATOGRAPHIC TECHNIQUES FOR PROTEIN SEPARATION AND PURIFICATION

G. M. de Plater
Division of Neuroscience and Division of Biochemistry and Molecular Biology, John Curtin School of Medical Research, Australian National University

INTRODUCTION

Chromatography is generally employed in neuroscience research for the purification of neuropeptides and proteins (e.g. neurofilament proteins) for further analyses requiring relatively homogenous material (e.g. structural analyses, pharmacology). Chromatography is also employed as an analytical technique, for both quantitative and qualitative determinations. Liquid chromatography[1] is the most commonly used chromatographic technique for protein purification, and other techniques will therefore not be discussed.

BASIC PRINCIPLES

Liquid chromatography involves three basic components: a solid stationary phase (column packing), a liquid mobile phase, and a column within which both phases are contained. The analyte (in this case a protein or peptide, designated A) is in equilibrium between the stationary (s) and mobile (m) phases as it moves along the column, according to:

$$A_m \rightleftharpoons A_s$$

The distribution of the analyte is described by the distribution coefficient, K_A:

$$K_A = \frac{[A]_s}{[A]_m}$$

K_A is a measure of the degree of retention. Retention is required to achieve separation (Figure 22.1). A more practical quantity than K_A is the capacity factor, k'_A, which can be determined directly from the chromatogram.

$$k'_A = \frac{\text{Total moles of A in stationary phase}}{\text{Total moles of A in mobile sphase}} = \frac{V_s}{V_m}\frac{[A]_s}{[A]_m} = \frac{V_s}{V_m} K_A$$

where V_s = volume of stationary phase within column
V_m = volume of mobile phase within column

The fundamental equation for the chromatographic process, which relates the retention volume, Vr, to other quantities, is

$$V_r = V_m(1+k'_A) = V_m + V_sK_A$$

V_r can be obtained directly from the chromatogram, as $V_r = Ft_r$,
where F = flow rate (ml min^{-1})
t_r = peak retention time

$$V_m = Ft_0$$

where t_0 = time required for a solvent molecule or other non-retained component to traverse the column (e.g. component C in Figure 22.1).

Substitution and rearrangement gives:

$$k'_A = \frac{t_r - t_0}{t_0}$$

INSTRUMENTATION

Liquid chromatography, in its simplest form, relies on gravity to draw liquid eluents through an upright column packed with a commercially available stationary phase. Fractions of a given volume are collected and analyzed after each run (usually by measuring their absorbance at a wavelength of 280 or 214 nm).

High-Performance Liquid Chromatography (HPLC) is an automated technique which relies on specialized equipment, consisting of an eluent delivery module, a sample application module, a column module, and an analyte detector module. The eluent delivery module is comprised of pneumatic or reciprocating piston pumps which pump eluent from reservoirs to the column. There are usually two pumps which allow the system to be run in gradient mode (i.e. where the composition of the eluent varies during a run, see below) where the delivery from each pump is computer-controlled. The sample is injected into the system via the application module, which ensures the complete airtight delivery of the sample to the column. HPLC columns are generally pre-packed with stationary phase and may be connected directly to the system. The analyte detector module is required for the continuous real-time monitoring of the mobile phase as it elutes from the column. For peptides and proteins the detector employed is generally a UV monitor set to record absorbance at a wavelength of 214 nm (peptides, small proteins with few aromatic amino acids) or 280 nm (proteins with aromatic amino acids).

Liquid chromatography techniques

The following is a brief description of the major chromatographic techniques used for protein purification.

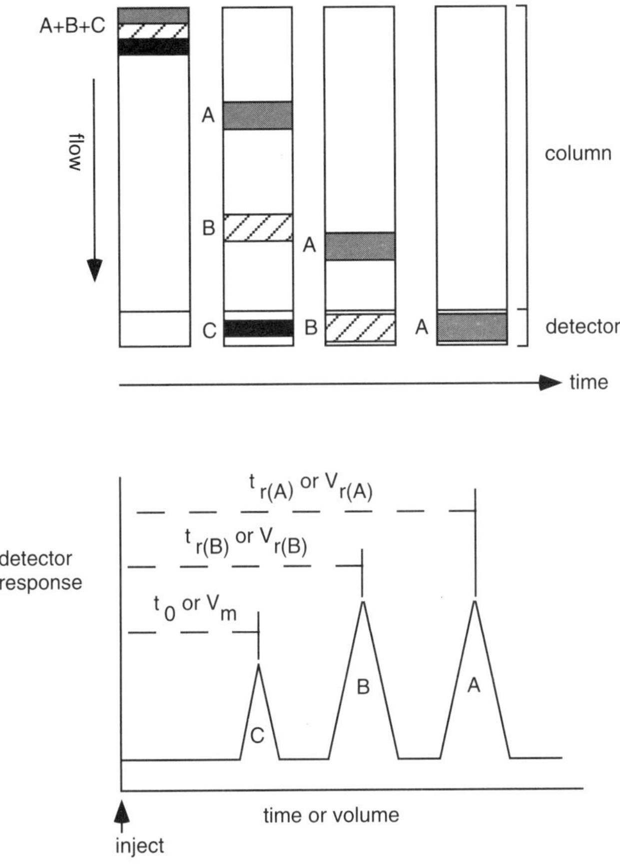

Figure 22.1 Diagram showing the chromatographic separation of three components: A, B and C, using a theoretical column to which component C cannot absorb (i.e. C is not retained). C therefore elutes at time = t_0 or volume = V_m. Components A and B are retained by the column to different extents and therefore elute at different time points.

Gel filtration

In this technique the stationary phase is either porous silica or a gel consisting of porous beads of cross-linked agarose. Retention time depends on the size of the solvated molecule in relation to pore size. Small molecules will traverse all or most pores whereas large molecules will be excluded by many (or all, if the molecule is larger than the pore exclusion limit). Larger molecules therefore have a smaller volume to traverse and hence elute earlier than smaller molecules. The retention volume, V_r is thus summarized by the following, where V_i = volume of column and V_t = total accessible volume:

$$V_r = V_m + K_A V_i$$

Substituting $V_t - V_m$ for V_i gives:

$$K_A = \frac{(V_r - V_m)}{(V_t - V_m)}$$

A calibration curve can be constructed by running a series of proteins of known molecular mass (M) and plotting logM versus K_A or V_r. Gel filtration can thus be used to determine the molecular masses of proteins under native (i.e. non-denaturing) conditions, in contrast to SDS-PAGE which relies upon analyte denaturation.

Ion exchange chromatography

Ion exchange chromatography exploits the fact that proteins and peptides exhibit a net positive charge when the pH of the mobile phase is greater than 1 pH unit below the isoelectric point (pI) of the protein/peptide and a net negative charge when the pH is greater than 1 unit above the pI.

Anion exchange chromatography involves a positively-charged stationary phase, e.g. quarternary amino ethyl (shown below), based on a normally hydrophilic polymer.

CH_3
/
–CH2–NH^+ quarternary amino ethyl
\
CH_3

The column is equilibrated with a buffer that is at least one pH unit above the protein/peptide of interest. The protein/peptide, having a net negative charge, adsorbs to the positively charged stationary phase and normally displaced by exchange with counterions from an increasing salt gradient. Retention and therefore the separation of negatively charged species is dependent on the relative strengths of their charges. Neutral or positively charged molecules are not retained by the column and elute in the void volume.

Cation exchange chromatography involves essentially the same processes, except that the stationary phase consists of negatively charged groups, e.g. sulphonate, based on a hydrophilic polymer.

$–CH_2–SO_3^-$ sulphonate

The pH of the mobile phase is chosen to be > 1 pH unit below the protein/peptide of interest. The adsorbed positively charged proteins/peptides are displaced by exchange with positively charged counter-ions from an increasing salt gradient. Again, separation is based on the relative strengths of the positive charges. Proteins that are neutral or negatively charged at the selected pH are not retained by the column and elute in the void volume.

Reversed phase chromatography

Reversed phase chromatography[2] (RPC) separates proteins and peptides on the basis of their hydrophobicity. The technique consists of a non-polar stationary phase, normally alkyl chains (e.g. octyl, C_8; octadodecyl, C_{18}) covalently bound to a base matrix (usually silica) to which proteins and peptides adsorb to a greater or lesser extent depending on their hydrophobicity. Ion-pairing agents such as TFA (trifluoroacetic acid) are included in the mobile phase to suppress charges on the protein/peptide and thereby enhance adsorption. Desorption takes place by decreasing the polarity (i.e. increasing the hydrophobicity) of the mobile phase by introducing an organic solvent (generally acetonitrile or isopropanol). Solvent molecules compete with the stationary phase for hydrophobic sites on the peptide/protein thereby effecting its displacement into the mobile phase, or more correctly, increasing K_A (i.e. shifting the $[A]_s \rightleftharpoons [A]_m$ equilibrium to the right). In RPC, the $A_s \rightleftharpoons A_m$ equilibrium is dynamic and as such peptides/proteins may be eluted by isocratic elution, i.e. a constant concentration of organic solvent as opposed to an increasing gradient. Generally RPC is applied to the separation of peptides and small proteins rather than larger proteins for two reasons: (i) the elution of large proteins is extremely sensitive to small variations in the concentration of organic solvent in the mobile phase and results are therefore less predictable and reproducible—a 1% change in the proportion of organic solvent can mean the difference between complete retention and almost immediate elution; and (ii) the presence of organic solvents in RPC can often lead to the irreversible denaturation of larger proteins.

Hydrophobic Interaction Chromatography

Hydrophobic Interaction Chromatography (HIC) is similar to RPC in that it separates proteins/peptides on the basis of their hydrophobicity. However, rather than using an organic solvent to elute the material from the hydrophobic stationary phase (which in HIC consists of a low density hydrophobic matrix such as phenyl superose), HIC employs a decreasing salt gradient. As hydrophobic interactions are enhanced by high salt concentrations, less hydrophobic proteins/peptides are eluted early in the gradient (i.e. high salt concentration) while more hydrophobic proteins elute later (i.e. low salt concentration).

Affinity Chromatography

Affinity Chromatography (AC) is a highly specific method for the separation of proteins/peptides on the basis of particular structural features. In AC, molecules which interact with the material of interest, such as a substrate or antibody are immobilized on a base matrix through which the protein/peptide sample is applied. Only the protein/peptide of interest should remain bound and all other material should elute in the void volume. Elution of the desired peptide/protein is effected by changing the mobile phase so as to dissociate the protein/peptide from its immobilized ligand. This is generally done by altering the pH or using an agent which competitively displaces the protein/peptide of interest.

RECENT PAPER

Scorpion (*Leiurus quinquestriatus hebreus*) venom contains a number of peptide toxins with potassium channel blocking activity. In order to investigate the activities of individual peptides, each must be purified to homogeneity from whole venom. To this end, Marshall et al. (1994) used several chromatographic techniques, starting with gel filtration and cation exchange chromatography. Each fraction from these steps was assayed for activity and the most active fractions were subjected to reversed phase HPLC as the final purification step. The resulting purified material is readily amenable to N-terminal amino acid sequencing, allowing Marshall et al. (1994). to not only compare activities but also the structures of individual toxins.

REFERENCES AND FURTHER READING

Kerlavage, A. R. (1989) *The Use of HPLC in Receptor Biochemistry*. New York: A. R. Liss.

Majors, R. E. (1989) Solid- and Liquid- Phase Chromatography. In *Instrumental Analysis*, edited by G. D. Christian and J. E. O'Rielly, (2nd edn). Massachusetts: Allyn and Bacon, Inc.

Marshall, D. L., Vatanpour, H., Harvey, A. L., Boyot, P., Pinkasfeld, S., Doljansky, Y., Bouet, F. and Menez, A. (1994) Neuromuscular effects of some potassium channel blocking toxins from the venom of the scorpion Leiurus quinquestriatus hebreus. *Toxicon* **32**, 1433-1443.

Oliver, R. W. A. (1989) *HPLC of Macromolecules: A Practical Approach*. Oxford: IRL Press.

NOTES

1. Liquid chromatography is commonly taken to mean liquid-solid chromatography rather than liquid-liquid chromatography, which involves a stationary liquid phase immiscible with the mobile liquid phase—the relative solubility of the analyte in each phase determines its K_A.
2. Reversed phase chromatography is so named because the phases are reversed relative to normal phase chromatography in which the stationary phase (e.g. unbonded silica) is more polar than the mobile phase (e.g. hexane). Polar compounds interact with the stationary phase and thus elute later than non-polar compounds, in contrast to RPC.

SECTION 9
PRODUCTION OF ANTIBODIES

CHAPTER 23

PRODUCTION AND USES OF POLYCLONAL ANTIBODIES

M. F. Crouch
Molecular Signalling, Division of Neuroscience, John Curtin School of Medical Research, Australian National University

INTRODUCTION

Antibodies are molecules which are produced by the immune systems of mammals and lower animals as part of the body's response to invasion by micro-organisms. Antibodies are made by lymphocytes and each antibody is relatively specific for a particular molecule or part of a molecule. For example, a bacterial infection elicits the production of a variety of antibodies in our bodies which will bind to distinct bacterial proteins such as those found on the bacterial cell wall. Each antibody is produced by a single cell and the body induces the division of these cells to increase the number of these specific antibody producing cells. These cells are thus clones of the original cell, and the process is called 'clonal expansion'. Once the antibody is bound to the invading organism, the antibody is then recognized by specialized cells in our body which destroy the antibody-bound microbe.

It is this specialized binding of antibodies to particular molecules which has made them such a useful tool to the cell biologist. As an antibody is molecule specific, if an animal is injected with a pure foreign molecule (a molecule not usually seen by the immune system of an animal), that animal will produce large amounts of antibody to that molecule. Therefore, blood taken from that animal during the phase that it is making these antibodies, will provide a source of these antibodies. They can be purified and separated from the other antibodies present in the blood. Then the antibodies can be used for biological studies, as discussed below.

HOW ARE SPECIFIC ANTIBODIES RAISED?

Rabbits are often used to raise antibodies to molecules of interest. The antibodies produced are called 'polyclonal' since although the antibodies of interest may be only to one molecule, many different antibodies may be raised to different parts of that molecule. Each antibody is produced by a single cell. Thus, as many different antibodies are produced to a single molecule, and these have come each from different cell clones, the collection of antibodies to a single molecule produced by an animal is termed 'polyclonal antibodies'.

To raise antibodies, a purified molecule, such as a protein, is injected into a rabbit. Alternatively, and commonly used these days, if the sequence of the protein of interest is known, peptides (short lengths of the protein) can be synthesized and used to induce antibody production. Use of such peptides overcomes the need to purify large quantities of the protein which may be in low abundance. In addition, it allows the easy purification of the antibodies, as discussed below.

Usually rabbits are injected several times over about 2–3 months, at intervals of about 3 weeks (Figure 23.1). For the production of anti-peptide antibodies, the initial injection is of about 0.5 mg peptide per rabbit. For purposes of enhancing the antibody production by the rabbit, the peptide is often coupled to a carrier protein that itself will raise an immune response, such as keyhole limpet haemocyanin (KLH, 2 mg) using glutaraldehyde (20 mM). The peptide, KLH and glutaraldehyde are dissolved or diluted in phosphate buffer. The initial injection of this peptide mixture is given with a 'complete' adjuvant, and subsequent boosts with 'incomplete' adjuvant. Half the amount of antigen and carrier protein is given in the boosts as with the initial immunization. Blood (about 30 ml) is collected from the rabbits by a small lesion made in one ear vein about 10 days after each booster injection. This can be repeated several times over the life of the animal to collect large quantities of antibody. The serum is separated from the red blood cells by inducing a clot, and the serum containing the antibodies can be kept frozen for long periods.

PURIFICATION OF ANTIBODIES

As serum contains not only the antibody of interest but also antibodies normally produced by the animal, it helps to purify the antibodies for biological studies so that there are fewer non-specific reactions to worry about. This is easily achieved with anti-peptide antibodies by coupling the synthetic peptide to an immovable support (such as a column, which is usually a glass tube filled with a porous substance to which molecules

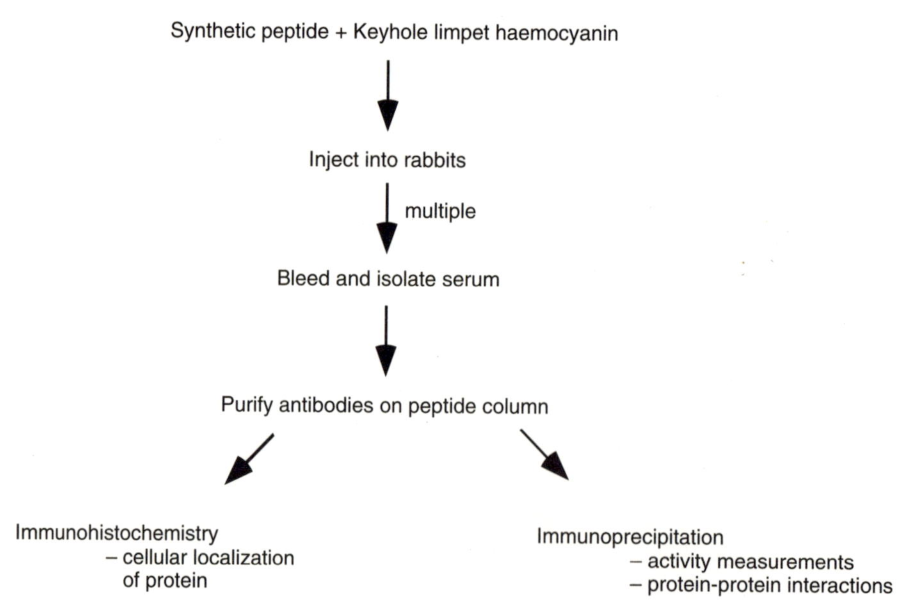

Figure 23.1 Basic procedure for production of rabbit anti-peptide polyclonal antibodies.

can be attached). When the anti-serum is passed over the immobilized peptide, it will bind to the peptide, but the other antibodies will not, and these will pass through the column. Those antibodies which are bound to the peptide can then be collected in a pure form by eluting them off the column usually with a low pH buffer and collecting them as they pass out of the glass tube. These antibodies can then be aliquoted into small volumes and frozen for long periods.

USES OF ANTIBODIES

Antibodies are extremely useful for examining the potential role of proteins in cell metabolism. For example, if a cell is activated by a hormone then the antibodies to that protein can be used to isolate the protein from inactive and hormone-activated cells; then it is possible to examine whether (a) the protein has become modified because of a reaction induced by hormonal activation, (b) the protein has been induced to link to other proteins, (c) the protein has changed its site in the cell. Cells can also be microinjected with antibodies to see if this affects a cellular process of interest.

As the antibody produced will bind specifically to the protein of interest, and theoretically not to others, the antibody can be incubated with cell extracts, such as from control and activated cells. The antibody will bind the protein and then another compound, called protein A-sepharose, can be used to bind the antibody. If the complex is centrifuged it will precipitate. Thus, it is possible to purify the antibody with its bound protein in a very short time. This process is termed 'immunoprecipitation'. The protein can be examined in detail by several means to determine whether it is modified or has other proteins bound to it.

A fluorescent tag can also be attached to the antibody and microscopic evaluation used to determine where the protein is in the cell. For instance, cells can be treated with a hormone, incubated with a fixative such that they are now dead but all the components of the cell are locked in their correct positions, and then incubated with a fluorescent antibody. Under the microscope the protein can be observed and its localisation compared with control cells. This method of using antibodies to examine proteins by microscopic means is called 'immunohistochemistry' (see Chapter 19).

FURTHER READING

Affinity Chromatography: Principles & Methods. Pharmacia.

Berven, L. A., Crouch, M. F., Katsis, F., Kemp, B. E., Harland, L. M. and Barritt, G. J. (1995) Evidence that the pertussis toxin-sensitive trimeric GTP-binding protein Gi2 is required for agonist- and store-activated Ca2+ inflow in hepatocytes. *J. Biol. Chem.* **270**, 25893–25897.

Crouch, M. F. (1991) Growth factor-induced cell division is paralleled by translocation of Gia to the nucleus. *FASEB J.* **5**, 200–206.

Goldsmith, P., Gierschik, P., Milligan, G., Unson, C. G., Vinitsky, R., Malech, H. L. and spiegel, A. M. (1987) Antibodies directed against synthetic peptides distinguish between GTP-binding proteins in neutrophil and brain. *J. Biol. Chem.* **262**, 14683–14688.

CHAPTER 24

PRODUCTION AND USES OF MONOCLONAL ANTIBODIES

W. D. Phillips
Institute for Biomedical Research, Department of Physiology, University of Sydney

INTRODUCTION

Antibodies are proteins of the immunoglobulin family each of which may have the potential to bind to a different antigen molecule. Antibodies are grouped into several immunoglobulin classes (e.g. IgG, IgM, IgE) and subclasses (e.g. IgG_1, IgG_2) that share common molecular structures and functional properties. Individual antibody members of an immunoglobulin subclass differ in their 'variable regions' (of the protein) and this difference makes them bind specifically to different target antigens. Antibodies are produced by the B-lymphocytes of the immune system. Each clone of B-lymphocytes (derived by mitosis from a single parent cell) is programmed to produce a single species of antibody with a single specificity for antigen.

Monoclonal antibodies (Mabs) can be produced by the fusion of an (immortalized) myeloma (cancer) cell line with a B-lymphocyte to produce a hybridoma cell line. The hybridoma clone is selected from among many for its production of antibody with the desired antigen specificity. Having isolated such a hybridoma it is possible to produce a virtually endless supply of highly specific antibody for research, diagnostic or other purposes.

Mabs are frequently used to detect, quantitate and localize proteins and other antigens within tissues. They can also be used to purify the antigen to permit biochemical analysis. Since the technology was introduced two decades ago (Köhler & Milstein, 1975), Mabs have become an integral tool in biomedical research and are playing increasingly important roles in clinical diagnosis and the biotechnology industry.

ADVANTAGES OF MONOCLONAL ANTIBODIES OVER CONVENTIONAL POLYCLONAL ANTIBODIES

Polyclonal antibodies are antibodies collected from the serum of animals (e.g. rats, mice, rabbits) that have been immunized. Unlike Mabs, they usually contain a mixture of antibody species that may recognize different substructures (epitopes) within the target antigen macromolecule. Unless they have been affinity purified, polyclonal antibodies may also be contaminated with antibodies that bind to other antigens. Thus Mabs offer the key advantage over polyclonal antibodies of specificity.

The other major advantage is quantity. The supply of polyclonal antibodies from an immunized animal is limited by the amount of serum (antiserum) that can be bled from the immunized animal. In contrast, the immortalized hybridoma cell line can, theoretically at least, provide an endless supply of the characterized antibody reagent. This is particularly important because the quality of antibodies produced following immunization can vary greatly between individual animals of the same species.

Mabs can be used for most of the common immunochemical techniques:

- immunoblotting (immunostaining antigens separated by gel electrophoresis)
- immunohistochemistry/immunocytochemistry (e.g. immunofluorescent or immunperoxidase staining of antigens in tissue sections or cultured cells)
- affinity purification of the antigen (for further analysis)
- radioimmunoassay and immunoprecipitation (used to quantitate antigen)

However, because they represent a single species of antibody with a single reactivity, Mabs need to be individually tested for their suitability in each of these tasks.

MAKING HYBRIDOMAS

Hybridoma cell lines are made by the physical fusion of cells of a myeloma cell line (such as the mouse myeloma line NS-1) with a mixed population of lymphocytes (usually dissociated from the spleen of an immunized mouse or rat). There are seven basic steps and these are outlined below and summarized in Figure 24.1.

1. Immunizing mice/rats

The aim is to immortalize a clone of B-lymphocytes that express an antibody specific for the antigen (protein, polysaccharide, neurotransmitter etc). By immunizing the animal with a highly purified preparation of the antigen (the immunogen), the proportion of B-cells of interest in the spleen should be strongly enhanced. Usually, animals are given at least two, and more usually three or more injections at 1–4 week intervals. Animals may be immunized with antigen purified from tissues, chemically synthesized peptides (attached to a larger carrier protein) or recombinant protein produced in bacteria. One recent development has been to 'immunize' mice by injecting into their skeletal muscle recombinant DNA containing a promoter and the coding sequence for the antigen protein. Such 'naked DNA' can be taken up and expressed by the muscle cells of the mouse or rat, leading to an immune response.

2. Cell Fusion

Cells dissociated from the spleen of the immunized mouse are mixed with cultured cells of an established myeloma cell line such as NS-1. A chemical agent such as polyethylene glycol (PEG) is added under conditions optimized to favor the fusion of the cells.

3. Cloning by limiting dilution

At least 500–1000 different clones of hybridoma cells (somatic hybrids between an antibody-producing B-cell and a myeloma cell) should be generated by the fusion. The clone or clones of interest must be physically isolated from other cells that either produce no antibody (and might overgrow the interesting clone) or produce an unwanted antibody. This is most commonly achieved by diluting the cell fusion mix and then distributing it into plastic 96-well tissue culture trays at such a density that, on average, there will be at most one viable hybridoma cell per well. It is usually necessary to provide a feeder layer of cells such as peritoneal macrophages (flushed from the peritoneal cavity of mice) to provide trophic support for the isolated hybridoma cell clone at these early stages, when the cell density is very low.

4. Drug Selection

To simplify the task of isolating a hybridoma, selective medium is used. This contains drugs such as aminopterin that prevent the growth of myeloma cells while permitting the growth of hybridoma cells that contain the genetic information from both the myeloma and the primary spleen cell.

5. Screening for antibody

Colonies of cells will develop in the bottom of the wells. Each colony arises from the division of a single hybrid cell produced by the PEG fusion. Some hybrid clones will produce no antibody, others will produce antibodies against antigens that may have contaminated the immunogen or against a virus to which the mouse may have been exposed. It is thus essential that before a fusion is undertaken for the first time a reliable, specific assay for the antibody of interest has been developed. The assay must be capable of quickly testing each of the 500–2000 clones likely to arise from a fusion. The hybridomas grow slowly at first but once they have reached about 50% confluence (cells cover 50% of the well bottom) they must be tested within about 24 hours so that decisions can be made about which clones will be subcultured. If left too long without subculturing, valuable clones may die out or be overgrown by other cells. Methods of screening vary from antigen to antigen and include enzyme-linked immunoabsorbant assay (ELISA), immunofluorescent staining of sections and immunoblotting. Where practicable, the method of screening should mirror the intended use of the antibodies. For example, if seeking antibodies to use for immunofluorescence on formalin-fixed tissue sections, this should be the method of screening. Often Mabs that bind well to denatured proteins on an immunoblot will fail to stain the same antigen on tissue sections.

6. Expanding clones of interest

Interesting clones are subcultured (passaged) into a larger volume of medium in a 24-well tray. When these wells become confluent, a sample of the cells can be frozen as a backup stock. To be quite certain that the interesting clone is not contaminated with some

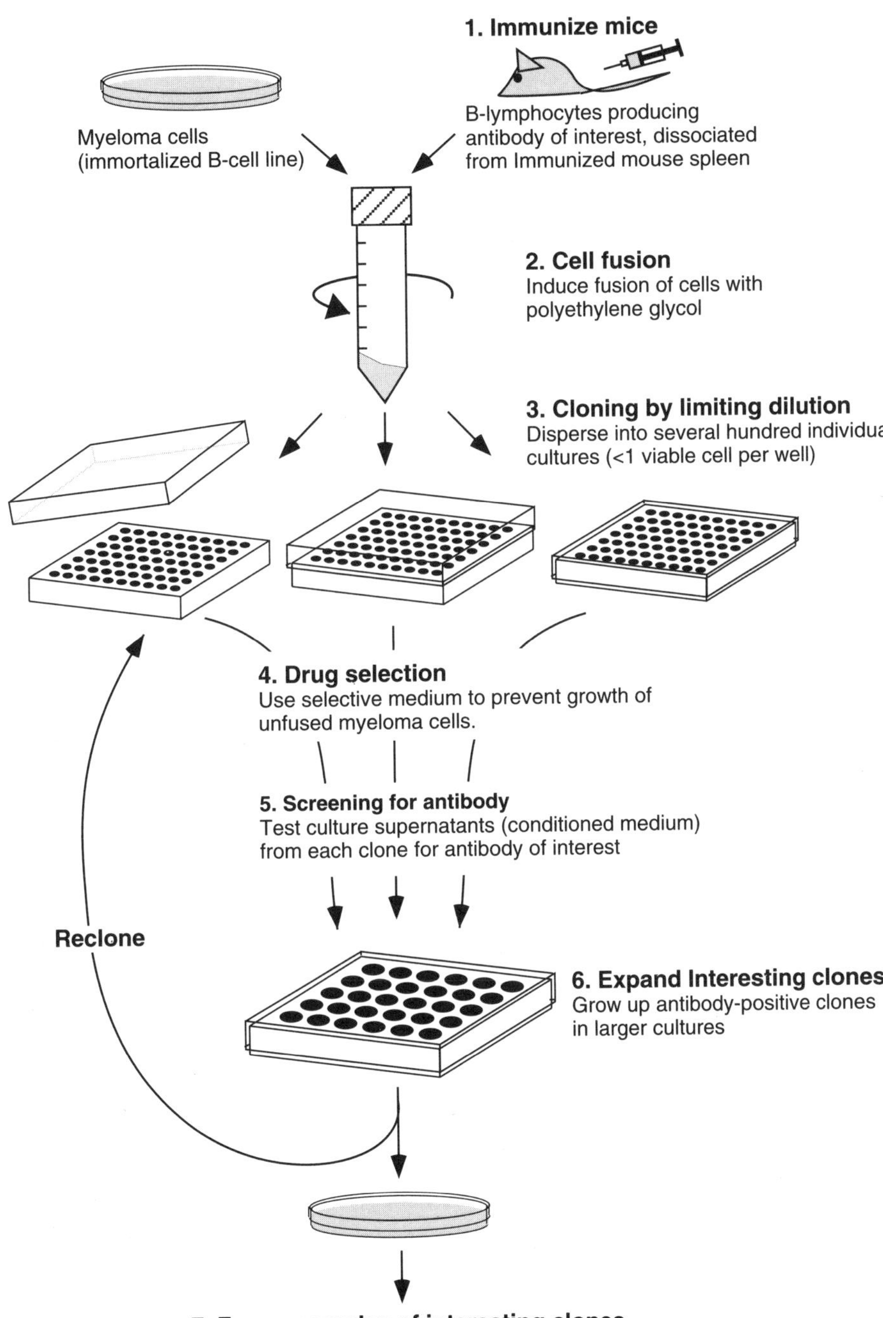

Figure 24.1 A general strategy for creating hydridoma cell lines to produce monoclonal antibodies.

other clone, it is advisable to repeat the cloning by limiting dilution (steps 3–5) at this stage. The clone can then be transferred into petri dishes and culture flasks so that larger numbers of antibody-producing cells (and larger quantities of antibody) can be produced.

'CLONE BY PHONE'

As with cDNA clones, it is often possible to obtain Mabs that have been isolated by other workers. A number of scientific journals and scientific societies have a policy that authors/members share clones and antibodies with other researchers to permit them to confirm results. This also facilitates the rate at which knowledge can grow. Unlike polyclonal antibodies, where the amount is limited, monoclonals can theoretically provide an endless supply of antibody, so people may be more generous in supplying you with samples. Monoclonals can come in several forms as illustrated in Table 24.1.

Table 24.1 Forms of Monoclonals

Form supplied	*Storage Conditions*	*Typical concentration*
Hybridoma cells	frozen in liquid nitrogen	need to grow up cells
Culture supernatant	frozen at –20°C, lyophilized (freeze dried) or liquid at 4°C	~10 μg/ml
Ascites fluid	frozen or liquid at 4°C	~ 10 mg/ml
Affinity purified	frozen or liquid at 4°C	variable

Some researchers lodge popular hybridoma clones with repositories such as the American Type Culture Collection (ATCC). The ATCC catalogue can be searched over the World Wide Web (http://www. atcc. org/catalogs. html). A wide range of Mabs are also available from various companies but you will pay dearly for them (e.g. A$200–500 for as little as 20 μg). This may make techniques that consume a large amounts of antibody, such as immunoprecipitation and immunoaffinity purification impractical. Commercially available listings such as the MSRS Catalog (Weimer, 1995) claim to provide a comprehensive inventory of all commercially available antibodies (polyclonal and monoclonal) to different antigens.

IMMUNOHISTOCHEMISTRY WITH MABS

Methods for immunofluorescence and immunohistochemistry with Mabs are essentially the same as those used with polyclonal antibodies (Harlow and Lane, 1988). Briefly, the primary antibody (specific monoclonal antibody) is first incubated with the cell or tissue section. Excess antibody that does not bind to the antigen is washed off. A secondary anti-immunoglobulin antibody (labelled with a fluorescent dye, histochemical enzyme or gold particle) is then incubated with the cell or tissue. The latter reagent should bind only to the monoclonal antibody which, in turn, is attached to the antigen *in situ*. Excess secondary antibody is then washed off and the distribution of the label (fluorescent dye, enzyme reaction product or gold particle) is visualized under a microscope. The appropriate concentration of a monoclonal antibody to use in incubation can only be determined empirically by testing several dilutions (10 μg/ml is a usual starting point for test dilutions). Control sections must be included in every experiment. In these, the primary antibody is deleted or replaced by some non-specific monoclonal to test the specificity of the interactions. In some cases it may be possible to use two different mouse monoclonals for double immunolabelling of two different antigens on a single section by using secondary antibodies specific to the particular IgG subtypes.

PITFALLS IN USING MABS

The key advantage of using monoclonal antibodies is their potential as a highly specific probe. A Mab should bind only to a single epitope (and possibly closely related molecular structures). However, an epitope can consist of as few as 5 amino acid residues of a protein or even just a sugar or lipid modification of a protein. In either case the epitope may be shared by one or more other macromolecules. Thus, it is necessary to exercise great caution in interpreting immunohistochemical staining. The presence of the antigen within the tissue should be followed up by immunoblotting, which permits the electrophoretic mobility of the antigen to be checked.

It is important to be aware that the effectiveness of a particular monoclonal antibody in a particular technique needs to be tested. For example, some Mabs will work well on immunoblots but fail to detect the antigen in tissue sections. Frequently a good polyclonal antibody will work better in immunoprecipitation assays than a monoclonal antibody against the same protein. In this regard, the very specificity of the antibody may be a disadvantage. Before setting out on making hybridomas for the first time it is important to consider carefully whether you might be better off raising polyclonal antibodies. Despite the fact that Mabs have been in wide use in neuroscience for fifteen years, leading research labs often opt to raise polyclonal antibodies in rabbits. The ease with which it is possible to produce large amounts of 'pure' antigen protein, in bacteria or by chemically synthesising a peptide sequence, can make the option of immunizing rabbits and affinity purifying the antibodies an attractive alternative.

With the increasing use of transgenic mice in neuroscience it is important to remember that it can be difficult to raise antibodies in the same species from which the immunogen was isolated or cloned. Thus problems may arise in trying to develop mouse monoclonals against rodent proteins. Furthermore, the presence of endogenous mouse immunoglobulins in sections of mouse tissue can provide a high level of background staining when using anti-mouse-immunoglobulin secondary antibodies. It is possible to produce rat or hamster hybridomas by fusing spleen cells from these animals with an established mouse or rat myeloma line but these hybridomas are sometimes unstable (see comments below).

Hybridomas are not always immortal producers of antibody. In some cases the hybrid cell may lose a chromosome, leading to the loss of antibody expression. This is especially pronounced with rat–mouse hybridomas. Free of the burden of producing antibody, such cells may quickly overgrow a hybridoma culture so that the antibody secreting clone is unrecoverable. Thus it is important to test the culture supernatants regularly and to keep frozen stocks of cells from the earliest times of culturing.

RECENT PAPER

Dystrophin-related protein (DRP/utrophin) is a recently cloned relative of the membrane skeleton protein, dystrophin. DRP is thought to play a role in stabilizing the postsynaptic membrane in skeletal muscle. Man et al. (1991) used part of the cloned gene to produce a fragment of the DRP polypeptide sequence in bacteria. By injecting this into mice they produced 19 monoclonals against DRP. Of these, 9 cross-reacted with dystrophin, highlighting the need to test monoclonals thoroughly for potential cross-reactivity. Nevertheless, by using the remaining Mabs, which variously recognized several distinct epitopes on DRP, they were able to overcome confusion in the literature about where these two proteins are expressed.

REFERENCES AND FURTHER READING

Harlow, E. and Lane, D. (1988) *Antibodies: A Laboratory Manual*. New York: Cold Spring Harbor Laboratory, Cold Spring Harbor.

Köhler, G. and Milstein, C. (1975) Continuous cultures of fused cells secreting antibody of predefined specificity. *Nature* **256**, 495–497.

Man, N., Ellis, J. M. Love, D. R. Davies, K. E. Gatter, K. C. Dickson, G. and Morris, G. E. (1991) Localization of the DMDL gene-encoded dystrophin-related protein using a panel of nineteen monoclonal antibodies: presence at neuromuscular junctions, in the sarcolemma of dystrophic skeletal muscle, in vascular and other smooth muscles, and in proliferating brain cell lines. *J. Cell. Biol.* **115**, 1695–1700.

Weimer, R. V. (1995) MSRS Catalog. Manufacturers Specifications and Reference Synopsis. Primary Antibodies. Birmingham, Michigan: Aerie Corporation.

SECTION 10

BLOTTING AND HYBRIDIZATION TECHNIQUES

CHAPTER 25

IMMUNOBLOTTING (WESTERN BLOTTING)

S. Johanson
Division of Neuroscience, John Curtin School of Medical Research, Australian National University

INTRODUCTION

Immunoblotting, also referred to as Western blotting, is a very useful and relatively simple technique used in many fields of medical research to determine the presence, quantity and molecular weight of solubilized proteins, and for assaying the quality and specificity of antibodies. In combination with immunoprecipitation (protein purification using antibodies), it enables sensitive detection of very low amounts of protein and can be used to investigate protein associations and certain protein properties such as phosphorylation. Major requirements for the technique are electrophoresis and immunoblotting apparatus, labelled secondary antibody and antibodies which recognize the proteins of interest (primary antibodies). This brief description of the basic concepts and procedures involved in immunoblotting is not intended as a complete instruction, for which other texts are available (see References and Further Reading).

Immunoblotting involves three basic steps:

1. separation of proteins according to molecular weight by gel electrophoresis;
2. transfer of separated proteins from the gel to a membrane support; and
3. immunodetection of the protein of interest.

Immunodetection involves, firstly, blocking of non-specific binding sites on the membrane. The protein of interest is then detected using labelled antibody, or unlabelled antibody followed by labelled secondary antibody. After detection, the protein of interest appears as a discreet band in a position corresponding to its molecular weight. Depending on the protein of interest and the antibodies used, amounts lower than 1 pg can be detected.

PROCEDURES

Gel Electrophoresis

Gel electrophoresis enables proteins in a sample to be separated according to their molecular weights. Strictly speaking 'immunoblotting' refers only to the transfer of proteins to a membrane support, followed by antibody detection, and does not include gel electrophoresis. However, gel electrophoresis is generally carried out prior to immunoblotting. The most common method of gel electrophoresis used when immunoblotting is

Tris/glycine sodium dodecyl sulphate polyacrylamide gel electrophoresis (SDS-PAGE). Protein samples (tissue or cell extracts, protein solutions, or immunoprecipitates) are prepared in sample buffer containing Tris buffer (a pH buffer), SDS (an anionic detergent which solubilizes proteins and gives them a negative charge), dithiothreitol (reduces proteins), and bromophenol blue (a blue dye). Cells or tissue samples can be prepared directly in sample buffer by homogenization or vigorous mixing, followed by centrifugation to remove residue. For protein solutions, a concentrated stock solution of sample buffer is simply added. Prior to electrophoresis, the samples are boiled for 5 minutes to fully denature the proteins and complex them with SDS.

Gel electrophoresis requires casting two polyacrylamide gels between glass plates separated by 0.75–2 mm. An upper stacking gel concentrates the samples before they enter the resolving gel, while the lower resolving gel separates the proteins (Figure 25.1). Some systems enable use of commercially available precast gels. Samples (usually 10–20 per gel) are loaded into wells in the upper stacking gel. Association of protein with SDS gives the protein a negative charge proportional to its mass, hence, when an electric current is passed through the gel in an electrophoresis apparatus, the proteins migrate down the gel towards the positive electrode (anode), forming separate lanes for each sample. The polyacrylamide gel serves as a molecular sieve which enables smaller proteins to move more readily through the gel than larger ones, thus leading to separation according to size. In most cases a protein's mobility is linearly proportional to the logarithm of its mass. Modifying the acrylamide concentration in the gel alters the pore size and hence the protein mobility; lower concentrations enable further migration and better separation of larger proteins. Ideally, prestained molecular weight standards (labelled proteins of known molecular weight visible on the gel and on the membrane after transfer) should also be run on the gel. Gels are usually run for 1–3 hours.

Transfer of proteins to a membrane support

Proteins are transferred from the gel to a membrane support by electrophoretic transfer in order to carry out immunolabelling. Nitrocellulose is the most commonly used form

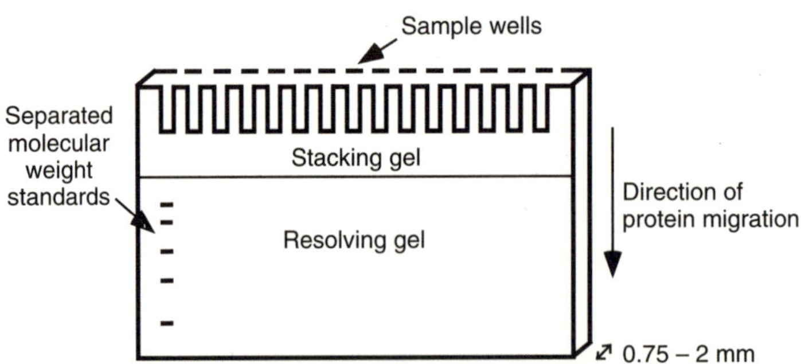

Figure 25.1 Polyacrylamide gel used in electrophoresis. The gel is cast between two glass plates (not shown).

of membrane, however polyvinyldiene fluoride (PVDF) membrane or others are also used. Transfer can be performed by one of two methods, both requiring specific transfer apparatus. Wet transfer involves placing the gel-nitrocellulose 'sandwich' (gel and nitrocellulose placed together between absorbent paper; Figure 25.2) into an assembly which is then immersed in a tank containing buffer solution and wire electrodes. Semi-dry transfer involves placing the sandwich (having soaked the absorbent paper in transfer buffer and nitrocellulose in water) between plate electrodes. In both cases, the electrical potential applied causes proteins to move towards the anode, thus the proteins migrate out of the gel and bind to the adjacent nitrocellulose. Wet transfer takes 4 to 18 hours while semi-dry transfer takes only 45 min to 1.5 hours, hence semi-dry is generally the method of choice. Wet transfer is best for efficient transfer of high molecular weight proteins.

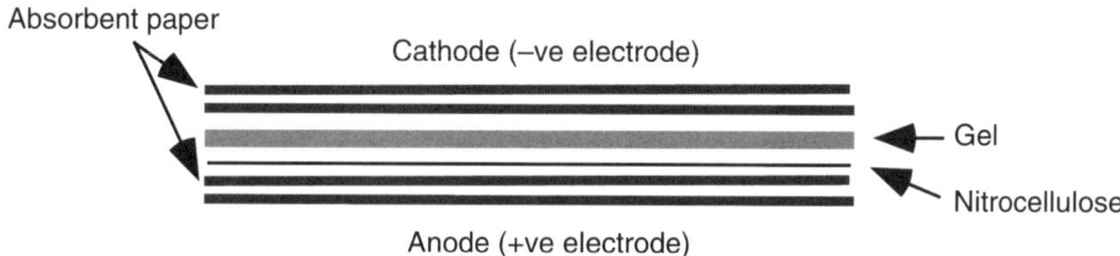

Figure 25.2 Assembly of the gel-nitrocellulose 'sandwich' for electrophoretic transfer of proteins.

Blocking non-specific sites on the membrane

The membrane (or blot) binds all proteins non-specifically, hence when transfer is complete it must be blocked to prevent non-specific binding of antibodies used in immunodetection. Blocking is performed by incubating the blot in a blocking solution, containing either 5% non-fat milk powder or 3% bovine serum albumin (BSA), in phosphate buffered saline solution (PBS), at room temperature for 1 hour. After blocking, the blot is washed prior to addition of primary antibody. PBS is commonly used for washing, however 0.1% BSA or 0.1% Tween 20 are often added to further reduce non-specific binding (Tween 20 is a detergent which itself binds to the blot and blocks non-specific binding sites).

Antibody incubation

Detecting the protein of interest on the blot is achieved using antibodies which bind specifically to the protein. These antibodies can often be purchased, depending on the antigen (protein of interest), or can be produced in the laboratory (see Chapters 23 and 24). The antibodies can themselves be labelled (e.g. radiolabelled) but are more commonly unlabelled and detected using labelled secondary antibodies. Labelled secondary antibodies are commercially available and enhance sensitivity over labelled primary antibodies, which generally must be labelled in the laboratory. Primary antibody is diluted in PBS solution containing 3% BSA or 0.1% Tween and incubated for at least 1 hr at room temperature. Incubations can be performed in dishes, heat-sealed bags, or more preferably, in cylinders which are rotated during incubation, in order to minimize antibody use. If the appropriate antibody concentration is unknown a range should be tried in order to determine the correct concentration, which varies widely between antibodies. Cross-reactivity of the primary antibody with proteins other than the protein of interest may be a problem and can sometimes be alleviated using lower antibody concentrations. Occasionally, antibodies which work effectively with other techniques are not suitable for immunoblotting.

After incubation with primary antibody the blot is washed thoroughly and incubated for 1 hour with a labelled secondary antibody, which binds specifically to the primary antibody. For example, if the primary antibody was produced in rabbit, labelled anti-rabbit Ig secondary antibody is used, which binds specifically to rabbit antibodies. The secondary antibody can be radiolabelled, but is more commonly labelled with enzymes such as horseradish peroxidase or alkaline phosphatase, which both catalyse substrate reactions enabling immediate detection. These enzyme-labelled secondary antibodies are available commercially. When using alkaline phosphatase labelled antibodies, dilutions and washes should be carried out using Tris-buffered saline rather than PBS. After secondary antibody incubation the blot is washed several times and the labelled antibody can then be detected.

Detection

Enzyme-labelled antibodies allow easy and rapid detection, however quantitation can be difficult due to the nature of the reactions involved. Several different enzyme substrates are available which all involve incubation of the blot in substrate for 1 to 30 minutes. A common choice is alkaline phosphatase using the substrate bromochloroindolyl phosphate/nitro blue tetrazolium (BCIP/NBT), which produces a black-purple precipitate with little background. Several horseradish peroxidase substrates can be used, such as chloronaphthol, aminoethylcarbazole or diaminobenzidine, of which the latter is the most sensitive, however the reaction is fast and can be easily overdeveloped. A more recent and increasingly popular method of detection involves the use of horseradish peroxidase labelled antibodies and a substrate reaction catalysed by the enzyme to produce luminescent light (chemiluminescence). This light is easily detected by exposure of the blot to autoradiography film. The technique is easy and fast and produces a stable copy of the result. Unlike other detection methods where the blot is permanently stained, the blot can be stripped of antibody and reprobed for other proteins. The method is particularly sensitive, however background staining can be a problem and quantitation can be more difficult than with other methods. Chemiluminescence substrate can be purchased in kit form with little preparation required.

It should be noted that varying methods to those described have often been used successfully and different antigens and/or antibodies might require variation from the standard methods.

IMMUNOBLOTTING IN NEUROSCIENCE RESEARCH

Immunoblotting is commonly used in neuroscience to assess the presence of particular proteins, to measure the relative amounts of proteins in different tissues or samples and to detect tyrosine phosphorylation of proteins, using phosphotyrosine antibodies. The technique is increasingly used to investigate the complex intracellular signalling pathways utilized by neurons. As an example, Ehlers et al. (1995) used immunoblotting to demonstrate that the receptor for nerve growth factor (NGF), a neuronal target tissue-derived growth factor acting on neurons, accumulates on both sides of a rat sciatic nerve ligation, hence is transported in both directions in nerve axons of the rat sciatic nerve. They further demonstrated that injection of NGF into the target (terminal) region caused an increase in the amount of receptor travelling back to the cell bodies from the terminals (retrograde transport). By immunoprecipitating (purifying with antibodies) NGF receptor from ligated sciatic nerve and immunoblotting these immunoprecipitates with antibody which binds only to phosphorylated tyrosine residues, they further showed that retrogradely transported NGF receptors are tyrosine phosphorylated and are therefore activated. This important finding demonstrates that retrograde transport of activated NGF receptor is likely to be important in the transduction of NGF signals from nerve terminals to distant neuronal cell bodies.

REFERENCES AND FURTHER READING

Bjerrum, O. J. and Heegaard, N. H. H. (1988) *Handbook of Immunoblotting of Proteins*, Vol. 1 and 2. Florida: CRC Press.

Ehlers, M. D., Kaplan, D. R., Price, D. L. and Koliatsos, V. E. (1995) NGF-stimulated retrograde transport of trkA in the mammalian nervous system. *J. Cell Biol.* **130**, 149–156.

Harlow E. and Lane D. (1988) Immunoblotting. In *Antibodies: A Laboratory Manual*, pp. 471–510. New York: Cold Spring Harbor Laboratory.

CHAPTER 26

SOUTHERN AND NORTHERN BLOTTING

S. Howitt
Division of Biochemistry and Molecular Biology, The Faculties, Australian National University

INTRODUCTION

Southern and Northern blotting are techniques used for the localization and quantification of particular DNA and RNA molecules, respectively. Both are specific applications of the filter hybridization technique, in which DNA or RNA is immobilized on a filter and a particular DNA or RNA molecule is identified by complementary base pairing to a labelled probe. What is unique to Southern and Northern blotting is that the DNA or RNA is first separated by gel electrophoresis and the transfer occurs from the gel to the filter. The technique was originally developed for DNA and named Southern blotting after its inventor (Southern, 1975). When the technique was adapted for use with RNA it became known as Northern blotting. Other applications of the filter hybridization technique include colony or plaque hybridizations, screening cDNA or genomic libraries and dot or slot blots.

METHODOLOGICAL APPROACH

Gel Electrophoresis

The first step in Southern or Northern blotting is separation of nucleic acids according to their size by gel electrophoresis (see Chapter 25). One of the major uses of a Southern blot is in identification of the genomic location and organization of a gene or multigene family. For these experiments, genomic DNA is isolated and digested with restriction enzymes before loading onto the gel. About 10 μg genomic DNA should be loaded for the detection of a single copy gene. The purpose of a Northern blot is often to determine the presence and abundance of a particular mRNA under certain conditions, e.g. in a particular tissue, at a particular developmental stage or in response to a defined stimulus. The source and quantity of RNA used will therefore depend on the particular application. An important control for this type of experiment is to analyze in parallel the expression of a gene which is not affected by the experimental treatment.

Membrane Transfer

Following gel electrophoresis, the next step is transfer of the DNA or RNA to a solid support, usually some kind of membrane. For a Southern blot, the gel is first washed in an alkaline solution for a short time to denature the DNA. The gel is then neutralized by washing in a buffer solution. Alkali treatment may be preceded by a depurination step (usually a wash in a mild acid solution) if the DNA molecules of interest are very large. Acid treatment results in the loss of purine bases from the DNA. During the subsequent alkali treatment, the DNA molecules will break at the depurinated sites, resulting in smaller molecules which transfer more efficiently. RNA is normally denatured with glyoxal plus dimethyl sulfoxide, methyl mercuric hydroxide or formaldehyde prior to electrophoresis.

A number of different membranes which bind DNA or RNA are available. Traditionally nitrocellulose was used and this still results in the lowest background. However, nitrocellulose membranes become very brittle and if the blot is to be re-probed, it may be better to use the more robust nylon membranes. Nylon membranes may be either neutral or positively charged, the latter having the advantage that transfer can take place under alkaline conditions, preventing the re-annealing of self-complementary sequences.

Figure 26.1 shows the set up for capillary transfer of nucleic acids to a membrane. This is the simplest and most commonly used procedure. The gel is placed on top of several sheets of filter paper which are bent over a support resting inside a tank of buffer solution. The ends of the filter paper are immersed in the buffer, forming a wick. Saranwrap is placed around, but not over, the gel to ensure that all capillary transfer goes through the gel. The membrane is cut to the same size as the gel and placed on top. This is overlaid with three more sheets of filter paper and a stack of paper towels which are weighted down. The paper towels and filter paper draw the buffer, and with it the nucleic acids from the gel, into the membrane by capillary action. Under these conditions, complete transfer takes 12–18 hours. Faster transfer may be obtained by using commercial systems for vacuum transfer or electroblotting. The final step of the transfer procedure is to fix the nucleic acid to the membrane. For nitrocellulose, baking at 80°C for two hours is used whereas a short exposure to UV light is required for uncharged nylon membranes.

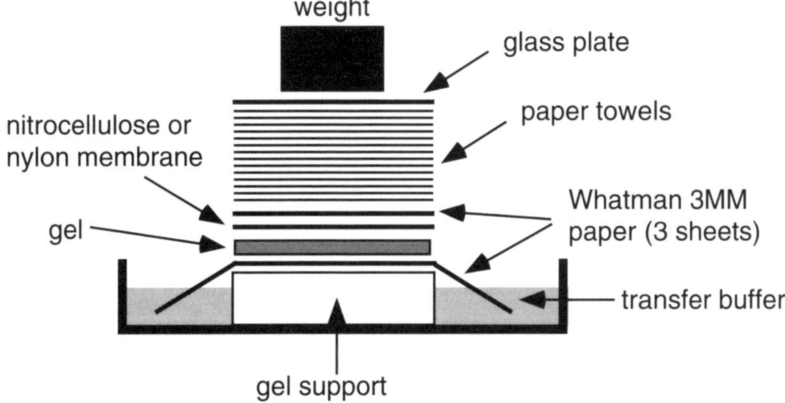

Figure 26.1 Capillary transfer of nucleic acid to a membrane.

Hybridization

Once the DNA or RNA is fixed to the membrane, it is ready for hybridization to the probe. Prior to hybridization a blocking reaction is performed in order to prevent non-specific binding of the probe to the membrane. The blocking solution should contain excess fragmented and denatured heterologous DNA. Alternatively, skim milk powder can be used but this generally results in a higher background. Either DNA or RNA can be used as a probe and the length can vary from an oligonucleotide of 15–20 bases to a fragment several kilobases long. The length and composition of the probe affect the degree of hybridization, with longer probes showing greater specificity and resulting in higher sensitivity (see Chapter 27). There are now a large number of different methods available for labelling and detecting the probe. Traditionally, radioactive labelling was used and this remains the method with the highest sensitivity. For the highest sensitivity, the probe should be labelled along its length, e.g. using nick translation or random priming. It is also possible to end label probes, resulting in a lower specific activity.

More recently, the options for non-radioactive probes have proliferated (see Kessler et al., (1995) for a full discussion). These methods depend on attaching some chemical group to the probe which can participate in a reaction resulting in the development of a colored product, chemiluminescence or fluorescence. One advantage of these methods is that labelled probe may be stored for a much longer time than radioactive probes. The choice of probe also depends on the sensitivity required, the availability of facilities for the use of radioactivity and the cost.

Conditions during the hybridization reaction and the subsequent washing steps are critical for determining the stringency of the reaction. Stringency affects the level of background obtained and the ability of the probe to bind to non-identical sequences. High stringency conditions are those in which only well-matched hybrids survive while low stringency allows the survival of mismatched hybrids. Lower stringency may be used to identify homologues of a particular gene in other species or members of a multigene family whereas high stringency is used for applications where it is necessary to distinguish between mutated and wild type sequences. Too high a stringency will prevent the formation of the desired hybrid while too low a stringency may result in the presence of false positives.

The critical factors for determining stringency in the hybridization reaction are the temperature and the composition of the hybridization solution. The determination of appropriate conditions is largely empirical. Annealing of complementary sequences is also dependent on the composition and length of the probe. The annealing temperature of a particular sequence can be calculated but the concentration of the target DNA is not known as this depends on the efficiency of transfer and the accessibility of the DNA on the membrane. In aqueous solutions, with a homologous probe, a temperature of 65°C is generally used. At this temperature nitrocellulose membranes become rather brittle and thermal denaturation of the probe can occur so an alternative is to include 30–50% formamide in the hybridization solution. Formamide lowers the annealing temperature, allowing the reaction to be done at 30–42°C.

Washing

Similar considerations apply to the conditions used for washing the blot. Washing is necessary to remove excess, unbound probe. Increasing the stringency of washes will reduce annealing of the probe to non-homologous sequences. Higher washing stringency can be achieved by a higher temperature, lower salt concentration or the presence of a low concentration of detergent. Thus the degree of hybridization can be controlled either by the hybridization conditions, the washing conditions or both.

Detection of the probe

The final step in the procedure is detection of the probe and the method used here will depend on the choice of probe. For a radioactive probe, this involves the exposure of the dried blot to X-ray film. For non-radioactive probes, some form of chemical reaction, producing a colored product, chemiluminescence or fluoresence is performed. For all methods, it is possible to quantify the amount of probe bound (which is proportional to the concentration of the target sequence) by densitometric analysis. It is also possible to strip the probe from the blot by various methods, allowing re-probing of the same blot for a different target sequence.

RECENT PAPER

A paper by Nedivi et al. (1993) describes the isolation and preliminary characterization of cDNAs which may play a role in long-term plasticity of the nervous system. A cDNA library was constructed from the dentate gyrus of rats after stimulation with kainate, which induces LTP-like potentiations as well as seizures. Subtractive hybridization and differential screening, involving the use of Southern blotting, was used to isolate 52 candidate plasticity-related genes. Northern blotting was used to confirm that expression of these genes was induced by kainate treatment. Control blots were done, probing for proenkephalin, which is induced by kainate and glycerol-3-phosphate dehydrogenase, which is not. The ratio of these two signals gives the level of kainate induction. Further experiments with Northern blots characterized the expression of some of the candidate genes under different conditions, e.g. tissue and developmental stage specificity, and the role of the NMDA receptor (which has been implicated in LTP) in induction.

REFERENCES AND FURTHER READING

Anderson, M. L . M. (1995) Hybridization strategy. In *Gene Probes 2*, edited by B. D. Hames and S. J. Higgins. Oxford: Oxford University Press.

Kessler, C. (1995) Non-radioactive labelling and detection of nucleic acid probes. In *Gene Probes 1*, edited by B. D. Hames and S. J. Higgins. Oxford: Oxford University Press.

Nedivi, E., Hevroni, D., Israeli, D. and Citri, Y. (1993) Numerous candidate plasticity-related genes revealed by differential cDNA cloning. *Nature* **363**, 718–721.

Sambrook, J., Fritsch, E. F. and Maniatis, T. (1989) *Molecular Cloning. A Laboratory Manual*, Chapters 7 and 9. New York: Cold Spring Harbour Laboratory Press.

Southern, E. M. (1975) Detection of specific sequences among DNA fragments separated by gel electrophoresis. *J. Mol. Biol.* **98**, 503–517.

Walker, P. D., Capodilupo, J. G. Wolf, W. A. and Carlock, L. R. (1996) Preprotachykinin and preproenkephalin mRNA expression within striatal subregions in response to altered serotonin transmission. *Brain Res.* **732**, 25–35.

CHAPTER 27

MEMBRANE AND IN SITU HYBRIDIZATION: PRINCIPLES AND PITFALLS

B. Key
Laboratory of Molecular Neurodevelopment, Department of Anatomy and Cell Biology, University of Melbourne

INTRODUCTION

Hybridization is the binding together of single strands of nucleic acids through hydrogen bonds (H-bonds) between apposing bases to form double strands. The specificity of this binding is determined by the degree of complementarity between bases in the strands; guanine binds to cytosine and adenine binds to thymidine or uridine. Membrane hybridization involves binding soluble nucleic acid strands to nucleic acid strands fixed to special paper membranes. In contrast, *in situ* hybridization involves binding nucleic acid strands to endogenous DNA or RNA in either tissue sections, cells in culture, cell smears or chromosome preparations.

The formation of double-stranded DNA is only possible because of the ability of nucleotide bases to form very specific intermolecular bonds. This bonding is referred to as Watson–Crick complementary base pairing. The discovery of the structure of DNA by the Nobel laureates Watson and Crick was facilitated by the realization that thymidine cross-linked very specifically to adenine while cytosine cross-linked to guanine. Guanine–cytosine base pairing involves three H-bonds and is hence stronger than thymidine–adenine base pairing which only involves two H-bonds. A single-stranded DNA molecule can thus act as a template to copy itself using complementary base pairing. The elongating chain is stabilized by formation of phosphodiester bonds between the sugar backbones of the bases. The end result is a helical structure stabilized by H-bonds between bases in adjacent strands, phosphodiester bonds between sugars and hydrophobic interactions between bases as they stack on top of themselves in each strand.

COMPLEMENTARY DNA STRANDS HYBRIDIZE

The discovery of Watson–Crick complementary base pairing set the stage for the subsequent technological revolution in molecular biology. The single most important technique involved the ability of single-strand DNA chains to hybridize very specifically with complementary strands to form double-stranded DNA. Hybridization is the foundation stone of molecular biology techniques. When a solution of heterogeneous double-stranded DNA chains is slowly heated the DNA chains will begin to denature and separate into individual strands as the H-bonds between complementary bases are broken. If double-stranded DNA is likened to Velcro strips then it becomes easy to imagine that it takes more energy to pull apart longer than shorter strips. Thus, the melting temperature (the temperature at which DNA denatures) will rise for longer DNA chains. Likewise, more energy is needed to pull apart Velcro which has more cross-links per unit length. Therefore a DNA chain with a higher content of guanine and cytosine bases per unit length will have a higher melting temperature since there are more H-bonds involved in base pairing. As the temperature is lowered below the melting temperature, DNA chains will begin to anneal and form stable double helices. If the temperature is lowered slowly the DNA chains will only anneal with their complementary strands within the mixture. Non-specific matches will be prevented from forming since their melting temperatures will be lower than that for perfect matches and hence the non-specific matches will denature. If the temperature is lowered quickly, then many mismatches will occur and the hybridization is said 'not to be stringent' or to have 'reduced stringency'. If the temperature is dropped quickly, to 4°C or below, than most DNA chains will remain as single strands since Brownian motion is minimal and there is not enough energy to form H-bonds between bases.

REACTION CONDITIONS GOVERN HYBRIDIZATION STRINGENCY

Hybridization can also occur between DNA strands of unequal length. The shorter the length of a DNA strand, the lower its melting temperature. There is, however, a trade-off between strand length and specificity of binding. As the strand becomes shorter there are more chances of the target sequence appearing at random in the genome and hence a greater likelihood that inappropriate hybridization will occur. Specificity is also reduced in long DNA probes if the hybridization temperature is too low. Low temperatures will facilitate hybridization between small contiguous sequences in the probe and complementary sequences in DNA. Only approximately 15 contiguous nucleotides need to be complementary in order for inappropriate hybridization to occur. However, if the temperature is raised to reduce mismatches then the hybridization is in danger of becoming too stringent because of the low melting temperature of the small probes. The shortest practical probe length at which effective hybridization can be performed is usually approximately 18 nucleotides. It is possible to manipulate the microenvironment during hybridization to facilitate perfect base pairing between DNA probes and target DNA. By performing the hybridization in high salt concentration it is possible to facilitate DNA–DNA interactions by a salting-out effect. As the sodium salt concentration rises, water molecules will interact preferentially with the charged sodium ions rather than DNA, which effectively causes DNA molecules to interact. Sodium ions also stabilize the double-stranded DNA by interacting with the negative charged phosphate groups and thus reduces electrostatic repulsion. Hybridization is also facilitated by addition of dextran sulfate or polyethylene glycol to the buffer since it reduces the effective water volume available to DNA and hence encourages nucleotide interactions. Hybridization stringency can be increased by addition of organic solvents such as formamide to the buffer. Formamide disrupts hydrophobic interactions between bases and DNA strands and hence effectively lowers the melting temperature. Thus, in the presence of formamide it is possible to lower the hybridization temperature and still maintain the same stringency as that achieved at higher temperatures without formamide. This can be used when higher temperatures are destructive to either membrane blots or tissues sections (see below).

HYBRIDIZATION USING RNA

Hybridization is not limited to base pairing between DNA molecules in solution. In fact, what has made hybridization so powerful in molecular biology is the ability to hybridize DNA probes to DNA which has been localized or fixed to solid substrates. Nuclear DNA can be isolated from cells, cut into small pieces with enzymes, and separated according to size using gel electrophoresis (see Chapter 25). The DNA fragments are then transferred on to specialized membranes which then can be hybridized with DNA probes in solution, a technique referred to as Southern blotting (see Chapter 26). By tagging the DNA probes with a radioactive or chemical marker it is possible to selectively visualize hybridized DNA on the membrane blots. Hybridizations can also be performed directly on chromosomes to localize genes on chromosomes via a technique referred to as fluorescence *in situ* hybridization, or FISH. In this technique DNA probes are labelled with fluorescent tags and hybridized DNA is visualized using reflected fluorescence microscopy.

Although hybridization technology initially exploited DNA–DNA interactions, it was not long before methods were developed to hybridize DNA to RNA, and RNA to RNA. RNA can be isolated from cells, separated by electrophoresis using special denaturing gels, and then blotted on to membranes, in a technique known as Northern blotting. The use of RNA probes, or riboprobes, only became possible with the advent of technology to synthesize these molecules *in vitro*. RNA polymerases were cloned and vectors containing RNA polymerase binding sites and transcription start sites were engineered which allowed both sense and antisense riboprobes to be generated. The sense riboprobes are identical to target RNA and serve as controls for specificity, whereas the antisense riboprobes are complementary to target RNA. RNA bases stack more closely than DNA bases within a helix and thus the increased stability provided by these hydrophobic interactions in RNA elevates the melting temperature of RNA–DNA and RNA–RNA hybrids in comparison to DNA–DNA hybrids of similar size. RNA probes are often favored because they allow more stringent hybridization conditions to be used. This is particularly important when probing a mix of DNA or RNA molecules that has many closely related sequences and only identical matches need to be identified.

RNA–RNA hybridization forms the basis for a powerful technique of identifying and quantifying low levels of mRNA, the ribonuclease protection assay. Labelled riboprobes are hybridized to mRNA in solution and any unbound single-stranded RNA is digested with ribonuclease. Remaining double-stranded RNA which is not degraded by this enzyme and hence is 'protected' is then electrophoretically separated and RNA levels quantified. This is a sensitive method for detecting low levels of mRNA in cells and tissues. However, this approach provides only an assessment of total mRNA from all cells analysed and provides little insight into regional or cell type differences in transcription of mRNAs.

DETECTION OF mRNA BY *IN SITU* HYBRIDIZATION

Often it is crucial to know which cells in a particular tissue are transcribing mRNA for a specific protein and to know how this mRNA is influenced by experimental manipulations. In these cases, the technique of *in situ* hybridization (meaning hybridization 'in place') enables the localization of mRNA at the single cell level. Tissue or cells are typically preserved with a cross-linking fixative such as paraformaldehyde which rapidly penetrates cells and immobilizes proteins and thus prevents widespread proteolysis and RNA degradation. Cells or tissue can then be processed for histological sectioning or, as often is the case for small embryos, treated as wholemounts and reacted en bloc. Cells are then permeabilized and protein surrounding mRNA removed by mild controlled proteolysis with exogenous proteinase K. Following removal of proteinase K by washing and re-fixation, mRNA is denatured to remove secondary structures by either heating or acid treatment. Subsequently, target mRNA is then hybridized with labelled riboprobes and visualized using a variety of microscopy techniques.

In situ hybridization is a powerful and reliable method for detecting low levels of mRNA at the resolution of a single cell. The type of probe used *in situ* hybridization is dependent on the level of sensitivity needed. For relatively abundant mRNA species such as house-keeping genes (eg. actin) it is sufficient to use small DNA oligonucleotides. The advantages of these probes are that they are easily synthesized and labelled at either their 5′ or 3′ ends and they readily penetrate tissue and cells. The disadvantage of these probes is that their small size lowers the specificity of base pairing and reduces the stringency of the hybridization. Also the number of labels that can be attached to the ends of these molecules is small so that their detection becomes limiting. Oligonucleotides are often used as probes for neuropeptides expressed by hypothalamic neurons in the central nervous system since these peptides are relatively abundant. Detection of membrane receptors such as olfactory receptor proteins, which are typically expressed at low levels, requires the use of larger riboprobes. One of the disadvantages of using larger probes is that there is increased non-specific binding, however, this is managed in the case of riboprobes by using a higher incubation temperature for hybridization. Large double-stranded DNA probes are generally avoided since the necessary stringency can not always be achieved and sensitivity is reduced because the anti-sense probe competes with both the sense strand and mRNA during hybridization.

LABELLED PROBES FOR HYBRIDIZATION

There are two types of labelled probes used in hybridization techniques: radioactive and non-radioactive. Radioactive-labelled probes are typically used for membrane hybridizations. The most commonly used radioactive molecule is ^{32}P since it is relatively simple to incorporate into nucleotides and has high energy which is beneficial in detecting low level signals. Radioactive labels are also commonly employed for *in situ* hybridization, however, their use often only reflects the past experience of researchers. These probes provide a diffuse signal due to wide scattering of radioactivity away from a point source. Consequently, resolution is not easily achieved at a single cell level in tissue sections.

Resolution can be enhanced by using ^{35}S and ^{3}H and more recently ^{33}P, however, the exposure times and technical difficulties of autoradiography using liquid emulsions is often prohibitive. Non-radioactive labels are now more commonly used for rapid and high resolution *in situ* hybridization. During synthesis of riboprobes, biotinylated or digoxigenin labelled nucleotides are incorporated into the elongating RNA. Following hybridization, antibodies against digoxigenin or biotin are used in standard immunohistochemical reactions to visualize sites of bound probe.

WHEN TO USE *IN SITU* HYBRIDIZATION

One of the problems with *in situ* hybridization is knowing when to use it. Often it is more important to know whether the protein is expressed then whether mRNA for the protein is present. There are examples in the literature (although not very common) where mRNA is present and yet no specific protein product is detectable. In some cases protein has been detected in whole tissue but levels have been too low to detect in individual cells. In this situation, *in situ* hybridization, which can detect as few as 30 transcripts per cell, is essential for localizing sites of expression. There are many proteins which are rapidly secreted from cells and *in situ* hybridization is needed to determine the source of expression in these tissues. Detection of mRNA by *in situ* hybridization is also used for identifying when inductive signals during development and differentiation of cells become active. However, the most common reason for using *in situ* hybridization rather than immunohistochemical tools is probably the relative speed with which localization information can be obtained by the two methods. Generally, it is more time efficient to generate probes for *in situ* hybridization, than it is to generate antibodies against protein products.

REFERENCES AND FURTHER READING

Bodkin, D. K. and Knudson, D. L. (1985) Assessment of sequence relatedness of double-stranded RNA genes by RNA-RNA blot hybridization. *J. Virol. Meth.* **10**, 45–52.

Emson, P. C. (1989) Hybridization histochemistry or in situ hybridization in the study of neuronal gene expression. *Comp. Biochem. Physiol.* **93A**, 233–239.

Meinkoth, J. and Wahl, G. (1984) Hybridization of nucleic acids immobilized on solid supports. *Anal. Biochem.* **138**, 267–284.

Puche, A. and Key, B. (1995) Identification of cells expressing galectin-1, a galactose binding receptor, in the rat olfactory system. *J. Comp. Neurol.* **357**, 513–523.

Vassar, R., Ngai, J. and Axel, R. (1993) Spatial segregation of odorant expression in the mammalian olfactory epithelium. *Cell* **74**, 309–318.

SECTION 11
EXPRESSION SYSTEMS

CHAPTER 28

CELLULAR EXPRESSION OF CLONED AND MUTATED ION CHANNELS

P. R. Schofield
Garvan Institute of Medical Research, Sydney

INTRODUCTION

The electrophysiological analysis of cloned and mutated ion channels requires the introduction of the DNA or RNA molecule encoding the ion channel of interest into a cellular host. This chapter will consider the relative merits and limitations of the use of the two most common methods of analysis of recombinant ion channels, microinjection and analysis in *Xenopus* oocytes and expression in mammalian cell lines.

EXPRESSION IN XENOPUS OOCYTES

Historically, the analysis of ion channels in heterologous systems has relied on the injection of messenger RNA (mRNA) into the oocyte of the African clawed toad, *Xenopus laevis*. Early studies demonstrated that the injection of total mRNA from brain or other tissues resulted in the formation of multi-subunit ligand-gated ion channel receptors with identical physiological and pharmacological properties to native receptors (Sumikawa et al., 1981). These demonstrations of the translational and post-translational fidelity of the oocyte system led to its widespread adoption as the method of choice for analysis of ion channels.

Many of the original studies concentrated on the voltage-clamp method to examine currents flowing in response to ligand activation. While almost all recombinant expression of ion channel proteins in oocytes now uses RNA transcripts generated by the use of the highly efficient bacteriophage SP6, T3 or T7 promoters, it is interesting to consider the early experiments by Mishina et al. (1984). To obtain mRNAs suitable for injection into oocytes, the cloned cDNAs were first expressed in monkey COS cells under the control of SV40 promoter and total mRNA then extracted for subsequent injection, expression and analysis in oocytes. Subsequently, Sakmann et al. (1985) demonstrated that single channel patch-clamp analysis was also easily accomplished in oocytes by using this system in their analysis of the contributions of specific receptor subunits to overall receptor function. Thus, the oocyte expression system has also been widely used for the structure/function analysis of mutated ion channel receptors with some of the most elegant studies being performed by Sakmann and Numa in their analyses of the ion permeation properties of the nicotinic acetylcholine receptor.

More recently, the *Xenopus* oocyte system has been used for the expression cloning of several receptors and ion channels including the substance K (tachykinin) receptor (Masu et al., 1987) and the glutamate GluR-K1 receptor (Hollmann et al., 1989). In the former instance, the method relies on second messenger (G-protein) coupling of the expressed receptor with an endogenous chloride channel of the oocyte while the latter method results in the screening of the directly expressed ion channels.

ADVANTAGES AND DISADVANTAGES OF THE OOCYTE EXPRESSION SYSTEM

In experiments involving expression cloning or analysis of mRNA populations extracted from tissues or cells, the larger physical size of the oocyte and therefore the ability to microinject tens of nanolitres of sample generated by *in vitro* manipulation continues to make them the expression host of choice. However, there are also a number of deficiencies and limitations in the use of the *Xenopus* oocyte system which has resulted in other expression systems being developed. This has included the somewhat laborious task of having to microinject all oocytes for expression studies, the difficulties of generating sufficient material for pharmacological analyses and the inability to use whole-cell preparations, for example in studies of channel assembly or for whole-cell patch-clamp recording.

EXPRESSION IN MAMMALIAN CELLS

Given the limitations of the *Xenopus* oocyte system, several groups sought to develop and validate mammalian cell expression techniques for the analysis of ion channels. Both transient and permanent expression systems had been developed and embraced by the molecular biology community in the early 1980s, although it wasn't until the mid to late 1980s when the desire to obtain enhanced expression systems for the analysis of cloned ion channels started to become a major limitation. Therefore, it was not suprising that existing mammalian expression systems were initially tested.

Claudio et al. (1987) used calcium phosphate mediated DNA uptake and selection for stable integration into host cell chromosomes, via the presence of a co-transfected selectable marker gene, to establish cell lines expressing the four subunits of the nicotinic acetylcholine receptor. In this study, thymidine kinase deficient murine fibroblast L cells were co-transfected with all four acetylcholine receptor cDNAs, which were under the control of the SV40 promoter/enhancer, and the thymidine kinase gene. The stable integration of all four receptor subunits was demonstrated by a combination of biochemical, pharmacological and electrophysiological techniques.

A similar approach, but using transient transfection methods, was developed by Pritchett et al. (1988) for the inhibitory $GABA_A$ and glycine receptor channels. In this instance the human embryonic kidney 293 cell line was used and receptor cDNAs were under the control of the human cytomegalovirus promoter/enhancer. Additionally, the modified calcium phosphate precipitation method developed by Chen and Okayama (1987) was also used to enable a much higher percentage of cells (between 10–40%) to be transfected. Transiently transfected cells require visual selection before electrophysiological analysis. In general, transfected cells can be distinguished by a fine calcium phosphate/DNA precipitate and, at least for $GABA_A$ and glycine receptors, some swelling of transfected cells also aids in selection.

As highlighted in both of the above studies, the great advantage of mammalian cell lines is that identical transfection protocols can be used to obtain receptor or ion channel preparations for ligand binding studies. Thus, since the electrophysiological and pharmacological data obtained are from the same system they are more directly comparable than data obtained from different experimental systems.

The use of transfected cells is not limited to any particular cell type. In fact, any cell in which recombinant ion channels can be expressed and does not express the channel of interest or have other channel activity resulting in an electrically noisy background is suitable for electrophysiological studies. Thus, one can find references to the use of COS, L, Hela and many other cell types in the literature, although the 293 cell line has generally been the preferred host.

ADVANTAGES AND DISADVANTAGES OF THE MAMMALIAN CELL EXPRESSION SYSTEM

Despite the advantages of mammalian cells for the combination of pharmacological and cell biological methodologies with the electrophysiological approaches outlined above, they still present difficulties in terms of selection of transfected cells. This can be overcome by the use of visually-selectable co-transfection markers such as fluorescent proteins or affinity tags which can be identified by treatment with suitably labelled monoclonal antibodies. One method that has been used by several groups is the co-expression of a cell surface marker such as the CD4 molecule and the simple visual detection of transfected cells by the use of anti-CD4 monoclonal antibodies coupled to colored or magnetized polystyrene beads.

MUTATION STRATEGIES

Structure/function analysis of cloned ion channel genes has been one of the more valuable approaches to defining important determinants of ligand binding, channel gating and ion permeation. Many mutation strategies having been developed and are readily available through commercial suppliers. While all involve the synthesis of one or a pair of overlapping oligonucleotide primers, most now offer high efficiency by the use of PCR-based methods or the utilisation of some form of two-step selection process. These mutation strategies can allow the insertion of restriction sites for the construction of chimeric molecules, direct PCR chimera construction or the generation of point mutants, insertions or deletions. Thus, the experimental approach is not limited by design or efficient generation of the desired mutant channel.

CURRENT TRENDS AND FUTURE PROSPECTS

Today, one finds ion channel research breakthroughs that use either *Xenopus* oocyte or mammalian cell expression systems. One can conclude from this that while both systems are valuable, there are certain experimental approaches which are most easily achieved via the use of one system, e.g. combination electrophysiological and pharmacological approaches using mammalian cells or electrophysiology based studies of ion channels using oocytes. In addition, certain practical limitations such as housing a *Xenopus* colony or familiarity with a given system means that each investigator will choose the method that is best suited to answering the particular problem being addressed. The student should remember that in many instances either method would be equally applicable.

Recent examples of this choice include the characterisation of the role of the β2 subunit of brain sodium channels in which co-expression of the β2 subunit with the a subunit results in increased functional expression, gating modulation and an increase in capacitance due to recruitment of membrane microvilli. This was demonstrated in *Xenopus* oocytes (Isom et al., 1995). The analysis of heritable mutations in the glycine receptor utilized a combination of pharmacological and electrophysiological techniques in transiently transfected 293 cells and demonstrated the role of a key residue in signal transduction and gating of the receptor (Rajendra et al., 1995). Often similar studies have been undertaken using either system as seen in the analysis of the surface accessibility of cysteine mutants of the voltage-dependent S4 domain of the sodium channel using transfected A201 cells (Yang & Horn, 1995) or the potassium channel using oocytes (Mannuzzu et al., 1996).

REFERENCES AND FURTHER READING

Claudio, T., Green, W. R., Hartmann, D. S., Hayden, D., Paulson, H. L., Sigsworth, F. J., Sine, S. M. and Swedlund, A. (1987) Genetic reconstitution of functional acetylcholine receptor channels in mouse fibroblasts. *Science* **238**, 1688–1694.

Chen, C. & Okayama, H. (1987) High-efficiency transformation of mammalian cells by plasmid DNA. *Mol. Cell. Biol.* **7**, 2745–2752

Hollmann, M., O'Shea-Greenfield, A., Rogers, S. W. and Heinemann, S. (1989) Cloning by functional expression of a member of the glutamate receptor family. *Nature* **342**, 643–648.

Isom, L. L., Ragsdale, D. S., De Jongh, K. S., Westenbroek, R. E., Reber, B. F., Scheuer, T. and Catterall, W. A. (1995) Structure and function of the beta 2 subunit of brain sodium channels, a transmembrane glycoprotein with a CAM motif. *Cell* **83**, 433–442.

Mannuzzu, L. M., Moronne, M. M. and Isacoff, E. Y. (1996) Direct physical measure of conformational rearrangement underlying potassium channel gating. *Science* **271**, 213–216.

Masu, Y., Nakayama, K., Tamaki, H., Harada, Y., Kuno, M. and Nakanishi, S. (1987) cDNA cloning of bovine substance-K receptor through oocyte expression system. *Nature* **329**, 836–838.

Mishina, M., Kurosaki, T., Tobimatsu, T., Morimoto, Y., Noda, M., Yamamoto, T., Terao, M., Lindstrom, J., Takahashi, T., Kuno, M. and Numa, S. (1984) Expression of functional acetylcholine receptor from cloned cDNAs. *Nature* **307**, 604–608.

Pritchett, D. B., Sontheimer, H., Gorman, C. M., Kettenmann, H., Seeburg, P. H. and Schofield, P. R. (1988) Transient expression shows ligand gating and allosteric potentiation of $GABA_A$ receptor subunits. *Science* **242**, 1306–1308.

Rajendra, S., Lynch, J. W., Pierce, K. D., French, C. R., Barry, P. H. and Schofield, P. R. (1995)

Mutation of a single amino acid in the human glycine receptor transforms β-alanine and taurine from agonists into competitive antagonists. *Neuron* **14**, 169–175.

Sakmann, B., Methfessel, C., Mishina, M., Takahashi, T., Takai, T., Kurasaki, M., Fukuda, K. and Numa, S. (1985) Role of acetylcholine receptor subunits in gating of the channel. *Nature* **318**, 538–543.

Sumikawa, K., Houghton, M., Emtage, J. S., Richards, B. M. and Barnard, E. A. (1981) Active multi-subunit ACh receptor assembled by translation of heterologous mRNA in *Xenopus* oocytes. *Nature* **292**, 862–864.

Yang, N. and Horn, R. (1995) Evidence for voltage-dependent S4 movement in sodium channels. *Neuron* **15**, 213–218.

CHAPTER 29

TRANSFECTION IN CELL CULTURES

W. D. Phillips
Institute for Biomedical Research, Department of Physiology, University of Sydney

INTRODUCTION

Transfection refers to techniques for introducing cloned, double-stranded DNA into cultures of dividing cells in such a way that it can be taken up by the nucleus. This makes it possible to express a cloned gene within cultured cells. Transfection can also be used to study the regulatory regions of a gene and for other purposes. Double-stranded DNA is typically mixed with a transfection reagent and added to the culture medium in which the cells are growing. A small part of the added DNA is then taken up by some of the cells and may be expressed. In the next few pages we will briefly consider the types of experiments that transfection makes possible, some common transfection methods and strategies and finally some alternatives to transfection.

USES FOR TRANSFECTION

Neuroscientists often use transfection to express proteins so as to study their functional properties (such as the gating properties and pharmacology of an ion channel). They may also use transfection to identify which parts of a gene are involved in regulating changes in its rate of transcription (e.g. changes in transcription rate in response to trophic factors or the electrical activity of the cell). Provided the cloned DNA and a suitable host cell are available, transfection can provide a great deal of information with much less expense and time than is involved in making transgenic mice.

There are three conceptual classes of experiments likely to be of interest to neuroscientists:

1. Expressing cloned proteins to study their functions

Proteins are commonly expressed in fibroblasts or other cell types to permit analysis of their biosynthesis and functional properties. For example, double-stranded complementary DNAs encoding the ion channel or receptor subunits of interest are ligated into a plasmid (a double-stranded circular DNA unit that can replicate in bacteria). The plasmid will contain a strong promoter element (the transcription initiating part of a gene). This is called an expression construct. The expression construct is then mixed with a transfection reagent and added to cultures of a cell line that does not normally express the channel/receptor in question. The complementary DNA is transcribed into mRNA, translated into protein and assembled as a functional ion channel/receptor on the cell surface. Voltage-clamp or patch-clamp analysis might then be used to study ion channels expressed in this way (for an example see Kopta & Steinbach, 1994). As well as studying the function of the wild-type protein, it is also possible to use *in vitro* mutagenesis of the complementary DNA to study the role of different amino acid residues in the overall function of the transfected protein (e.g. Kanai & Hirokawa, 1995).

2. Studying the transcription-controlling regions of genes

The parts of a gene that are important in regulating its transcription (i.e. determining which cells express it and how this expression is modulated during life) can be identified and studied by joining (ligating) different portions of the gene (often from the 5′ side of the transcription start site) into a plasmid containing a 'reporter gene'. This reporter construct is then transfected into cultured cells. The reporter gene is usually an enzyme marker like luciferase, chloramphenicol acetyl transferase or β-galactosidase that produces an easily assayed enzymatic reaction product. The main idea in these sorts of experiments is to see how effectively the cloned genomic fragment serves to regulate transcription of the reporter gene (and by inference, the endogenous gene). By deleting or mutating portions of the putative promoter and/or enhancer DNA, it is possible to test which nucleotide sequences are important for the patterns of expression peculiar to a given gene (for an example see Chu et al., 1995). In these studies it is essential to use, as the host, a cell type that mimics the pattern of regulation of the endogenous gene in the tissue concerned.

3. Analyzing RNA sequences involved in targeting or stabilizing messenger RNA (mRNA) within the cell

This is a recent area of interest arising from evidence that sequences within the 3′ untranslated region of the mRNA can be important in determining the metabolic stability and subcellular destination (e.g. dendrites versus cell soma) of certain mRNAs. Cells are transfected with expression plasmids containing 3′ untranslated sequences from a particular mRNA ligated to sequences that serve as a tag or marker for the hybrid mRNA. Mutated forms of the same hybrid mRNA can be made by *in vitro* techniques (see Chapter 28) to test the importance of specific sequences within the 3′ untranslated region in determining the stability and/or subcellular location of the mRNA (see for example Behar et al. 1995).

METHODS OF TRANSFECTION

Expression constructs

The cloned complementary DNA encoding a protein of interest is usually ligated into an existing expression plasmid (one that has been successfully used to express other proteins). An expression construct made in this way (Figure 29.1) can then be transfected into the cultured cells using any one of the several transfection agents outlined below.

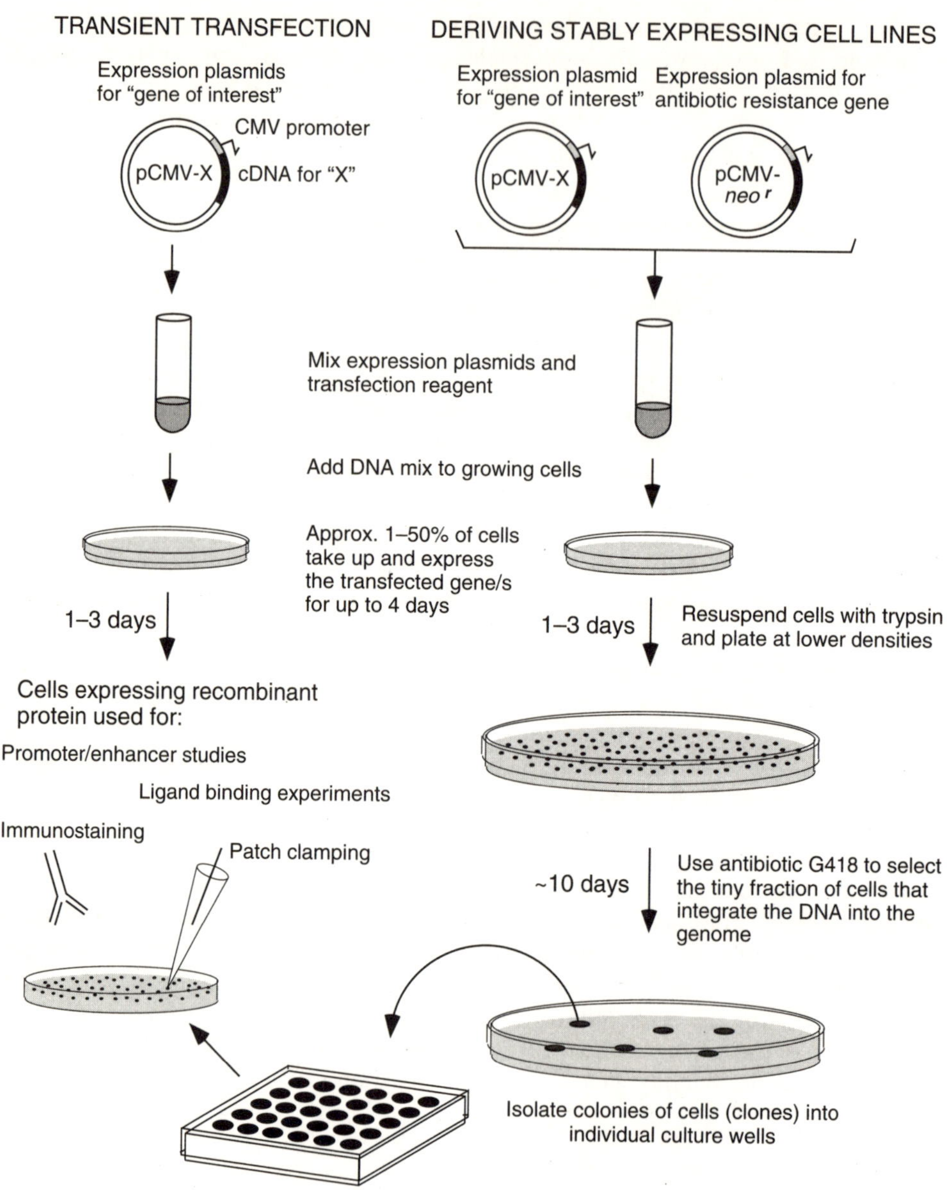

Figure 29.1 Transfection of mammalian cell lines for short-term experiments (transient assays) and for making stably expressing cell lines.

A number of expression plasmids have been constructed by various researchers and biotechnology companies around the world. In general these contain a promoter to permit initiation of transcription in mammalian cells and a restriction site (or a group of sites known as a polylinker) into which your complementary DNA can be inserted. An expression plasmid may also contain other sequences that help to ensure that the transcribed mRNA is efficiently processed and exported to the cytoplasm for translation. Choosing an expression plasmid that has been successfully used to express other proteins may increase your chances of obtaining good expression. It is important to consider the type of promoter. Most commonly used are the viral promoters. The human Cytomegalovirus Immediate Early gene promoter (often abbreviated CMV) and Rous Sarcoma Virus long terminal repeat promoter (RSV-LTR) are perhaps the two most widely used. Both produce high level expression of the gene of interest in a wide range of cell types. The β-actin promoter forms the basis of some expression constructs and these should also ensure robust expression in a wide range of mammalian cell types. In addition, tissue-specific promoter/enhancer constructs may be obtained from individual researchers. Different expression constructs can work with varying efficiencies depending on the retinue of transcription factor proteins in the cell type transfected. It is important to confirm from the literature or, if necessary, by pilot experiments that a given promoter will function within the cell type you intend to transfect.

The ligation of the cDNA into the expression plasmid will involve some basic DNA subcloning techniques (Maniatis et al., 1989). If you do not have ready access to the cloned cDNA but know the full sequence, it is possible to use reverse transcription followed by polymerase chain reaction to amplify and isolate the complementary DNA direct from the source tissue. This is most easily achieved for small proteins (say less than 50 kiloDaltons). Polymerase chain reaction becomes more problematic for larger complementary DNAs.

Transfection agents

1. Calcium phosphate transfection

This involves mixing the expression plasmid together with calcium chloride and phosphate buffer solutions. The calcium and phosphate ions form an insoluble precipitate in which the plasmid DNA becomes trapped. Tiny crystals become attached to the surface of the cultured cells. The DNA is then taken up by the cells and into the nucleus by a poorly defined mechanism. This leads to transcription and translation of the gene of interest. Calcium phosphate transfection is a classic technique and with careful application of the refined protocol (Maniatis et al., 1989) it is cheap and can be quite reliable. The transfection efficiency (defined as the percentage of cells within a transfected culture that express the transfected gene) can range from less than 1% (e.g. with primary muscle cells) to as much as 80% (with some fibroblast cell lines) depending upon cell type and upon the optimization of the technique. Particular attention needs to be applied to the pH of the phosphate buffer solution and to empirically determining the correct amount of plasmid DNA per transfection mix.

2. Transfection with liposomes

A range of these preparations are commercially available and can be tested upon your own particular cell type. They are claimed to be gentler on the cells than calcium phosphate transfection and may permit at least some level of transfection in cell types where calcium phosphate totally fails. In our hands with fibroblast and muscle cell lines, both calcium phosphate, and DOTAP (a liposome preparation from Boehringer Manheim) have been successfully used. Primary cultures of cells are considered difficult to transfect but we have obtained transfection efficiencies of up to 1% (of the cells) with Lipofectamine (Life Technologies, USA).

3. DEAE-dextran

The expression plasmid DNA is mixed with a commercially available DEAE-dextran and added to the cells. As with liposomes, DEAE-dextran transfection should provide a more consistent transfection efficiency. It is claimed that cells transfected in this way are unsuitable for deriving stably expressing cell lines. As with the above agents, the efficacy of DEAE-dextran must be determined empirically for the cell type in question.

4. Electroporation

Some cell types are particularly resistant to transfection by all the above methods. This is often the case with primary cell cultures (cells isolated directly from tissue rather than immortalized cell lines). In these cases it may be necessary to resort to electroporating your expression plasmid into the cells. Electroporators are available from several of the major biotechnology companies. They apply a brief, high voltage across a suspension of cells by rapidly discharging a capacitor. Adherent cells need to be first resuspended in a low salt solution. Application of the electric field produces many tiny pores in the cells through which the plasmid can enter the cell. Low salt conditions are necessary to avoid electrical arcing. Electroporation offers a high efficiency of transfection at the expense that about half the cells in the suspension will be killed under optimal conditions.

Transient transfection

Following transfection by one of the methods outlined above, some fraction of the cells in the transfected culture should express the transfected protein for a period of 3–4 days. The length of time that the protein can be detected in the cells will depend upon the rate of division of the cells (the plasmid will be 'diluted out') as well as the metabolic stability of the mRNA and of the protein itself. In our hands, protein expression can be detected at the earliest times tested (5 hours after transfection).

Important considerations and pitfalls in using transient transfections:

- Cells must be actively dividing in order to take up and express expression plasmids.
- Very high level of expression (much higher than the normal (endogenous) levels) can be expected during the first few days when using viral promoters.
- The level of expression can vary greatly between cells in the same transiently transfected culture.
- Two or more expression plasmids can be mixed together and co-transfected. Provided that the same promoter is used for each expression plasmid co-transfection will typically result in co-expression of each of the genes in the mixture (i.e. perhaps 80%

of cells that express one protein also express the others). This is sometimes not true for stably transfected cell lines. We have transiently co-expressed as many as 5 different proteins, making it possible to assemble multi-subunit ion channels and study protein–protein interactions in the living cells.

- Transfection efficiency can be very variable. Thus, for any kind of quantitative study it is important to co-transfect a control plasmid such as an expression plasmid for β-galactosidase. The level of expression of β-galactosidase can then be used to compare transfection efficiencies between different transfections performed on a given day.
- The low efficiency of transfection can be a problem in patch-clamping studies where it might be necessary to patch 10 or more cells before finding one with the transfected ion channel. A major technical development has been the recent availability of expression plasmids for jellyfish Green Fluorescent Protein (GFP) and its derivatives. These expression plasmids are commercially available from a number of sources and can be co-transfected with the ion channel expression plasmid/s. The GFP fluoresces in the living cell under short wavelength excitation light, allowing the minority of transfected cells to be quickly identified. There are a number of derivatives of the GFP that differ greatly in the brightness with which they fluoresce, so one should choose with care.

The basic steps involved in a transient transfection experiment are outlined in Figure 29.1 (left hand side).

Stable transfections

If cells are co-transfected with expression plasmids containing the gene of interest together with and an expression construct conferring drug resistance (e.g. the neomycin resistance gene, *neo*^{r}), a small proportion of the transfected cells will integrate the transfected DNAs at seemingly random site/s within their chromosomes. The expression construct for *neo*^{r} confers resistance to the antibiotic, G418, and is used as a selectable marker. That is, when G418 is added to the medium, only those cells that have stably incorporated the *neo*^{r} gene into their genome can survive and form clones of cells. A considerable proportion of G418-resistant clones are typically found to stably co-express the co-transfected gene of interest. As an alternative to co-transfection, it is also possible to construct a plasmid that contains expression constructs for both *neo*^{r} and the gene of interest. In this case an even higher proportion of G418-resistant clones will express the gene of interest. The steps involved in preparing a cell line that stably expresses your gene of interest are illustrated in Figure 29.1 (right hand side).

The following features and pitfalls of stably transfected cells should be considered:

- It is impossible to predict where in the genome the transfected DNA will become incorporated but often multiple copies become inserted at the a single site.
- The site of insertion in the genome can drastically affect the level of expression of the gene of interest, irrespective of the promoter used. There may also be unexpected mosaic expression, with only a portion of cells within a clone expressing the gene/s of interest.

- When using cell lines 'stably expressing' two or more transfected genes it cannot be assumed that all the cells within the culture are co-expressing each of the genes simultaneously. For example, some cells within a clonally-derived culture may express protein 1 but not protein 2 or vice versa at any given time. This is important when one wishes to infer something about protein–protein interactions or about the contributions of different subunits to ion channel properties. Co-expression must be confirmed at the single cell level by immunofluorescence or some other means.
- While the level of expression of a stably expressed gene can vary greatly from cell line to cell line, it is typically much lower than what is seen in the first few days of a transient transfection experiment.
- Comparison between different cell lines, even if derived from the same parental cell line, must be undertaken with great caution as the process of transfection and cloning may isolate characteristics unrelated to the effects of the gene of interest. It is usually necessary to isolate and study a number of clonal cell lines to confirm that the changed properties of the cell are consistently associated with the presence of the gene of interest.

Alternatives to transfection

Injection of *Xenopus* oocytes

Cloned ion channels are often expressed in frog (*Xenopus*) oocytes to facilitate voltage-clamping analyzes (see Valera et al., 1994 for a recent example of the use of this approach). This is achieved by injecting synthetic (recombinant) mRNAs into the oocyte. This technique can be very useful for voltage-clamp studies where a reliable supply of *Xenopus* oocytes has been established. Key advantages are the large size of the oocytes which facilitates patch-clamping, ability to readily identify cells that have been injected (and therefore are likely to be expressing the ion channel) and the ability to control the amount of the mRNA injected (and hopefully the amount of the receptor subunits expressed). A major disadvantage is the need to maintain a supply of *Xenopus* oocytes in the laboratory. The amount of recombinant protein produced by injected oocytes is severely limited, making some biochemical and pharmacological studies impractical. In contrast transfection offers the advantages that it is technically straightforward (given some experience in cell culture), both short-term (1–4 days) and long-term (up to months) expression can be studied, and large numbers of expressing cells can easily be produced, permitting biochemical and pharmacological studies of the expressed protein/s.

Viral vectors

Some cells may be highly resistant to all of the common transfection techniques. In this case viral vectors may be a last resort. While adenovirus and herpes viruses are beginning to be used as vectors for these purposes (particularly for their ability to infect postmitotic cells like neurons), retroviruses are still the best established and most convenient viral vector. Retroviral constructs can be made fairly easily by routine DNA subcloning. The modified viral genome (containing the gene of interest) can be packaged into the virion particle by simply transfecting the DNA into an established packaging cell line.

The culture supernatant of the packaging cell line (which will contain the packaged recombinant virus RNA genome) can then be used for infecting the chosen host cells. A reverse-transcribed copy of the RNA genome of the retrovirus becomes incorporated into the genome of the infected cells permitting stable expression.

RECENT PAPERS

Over a period of time colleagues have used transgenic mice to identify a fragment of the gene for the epsilon subunit of the acetylcholine receptor that contained the necessary information to ensure the gene was transcribed just by the sub-synaptic nuclei of skeletal muscle fibers. In 1996 they transfected reporter constructs into the muscle cell line Sol-8 to test whether the nerve-derived signalling protein, ARIA, might act to regulate the epsilon subunit gene via this fragment of the gene (Chu et al., 1996).

The microtubule-associated proteins, MAP2 and Tau become differentially concentrated during development in the dendrites and axons of the neuron respectively. Several cell-biological processes, including subcellular targeting of the mRNAs, contribute to this localization. Kanai and Hirokawa (1995) transfected expression constructs encoding either MAP2 or Tau into cell cultures from the spinal cord and localized the transfected proteins after neurons become differentiated. Various mutant forms of these proteins were transfected to determine which parts of the protein are important for determining its selective location within the neuron. They attached an 'epitope tag' to cDNAs to allow the recombinant protein to be distinguished from the endogenous protein.

To study the contributions made by specific subunits to the channel properties of the nicotinic acetylcholine receptor, expression plasmids for the four subunits that make up the fetal (α, β, γ, δ) or adult (α, β, ε and δ) acetylcholine receptor were co-transfected together with the *neo*^{r} into a fibroblast cell line (QT-6) by Kopta and Steinbach (1994). Clones of cells resistant to the antibiotic G418 (conferred by *neo*^{r}) were isolated and shown to synthesize and assemble the subunits into a functional acetylcholine-gated channel. This approach enabled them to relate the channel gating properties and pharmacology to the subunit composition.

REFERENCES AND FURTHER READING

Behar, L., Marx, R. Sadot, E. Barg, J. and Ginzburg, I. (1995) Cis-acting signals and trans-acting proteins are involved in tau mRNA targeting into neurites of differentiating neuronal cells. *Int. J. Devl Neurosci.* **13**, 113–127

Chu, G. C., Moscoso, L. M. Sliwkowski, M. X. and Merlie, J. P. (1995) Regulation of the acetylcholine receptor ε subunit gene by recombinant ARIA: an *in vitro* model for transynaptic gene regulation. *Neuron* **14**, 329–339.

Kanai, Y. and Hirokawa, N. (1995) Sorting mechanisms of Tau and MAP2 in neurons: suppressed axonal transit of MAP2 and locally regulated microtubule binding. *Neuron* **14**, 421–432.

Kopta, C. and Steinbach, J. H. (1994) Comparison of mammalian adult and fetal nicotinic acetylcholine receptors stably expressed in fibroblasts. *J. Neurosci.* **14**, 3922–3933.

Maniatis, T., Fritsch, E. F. and Sambrook J. (1989) *Molecular Cloning. A Laboratory Manual*. New York: Cold Spring Harbor Laboratory Press.

Valera, S., Hussy, N. Evans, R. J. Adami, N. North, R. A. Surprenant, A. and Buell, G. (1994) A new class of ligand-gated ion channels defined by P_{2x} receptor for extracellular ATP. *Nature* **371**, 516–519.

SECTION 12
GENETIC TECHNIQUES

CHAPTER 30

REVERSE TRANSCRIPTION POLYMERASE CHAIN REACTION

M. Vidovic
Developmental Neurobiology, Research School of Biological Sciences, Australian National University

INTRODUCTION

The polymerase chain reaction (PCR) technique was developed to amplify, with high efficiency, double-stranded or single-stranded DNA sequences. Amplification of individual RNA molecules can also be accomplished with PCR when coupled with reverse transcription (RT) of RNA into a complementary DNA (cDNA) copy. The method that combines RT-PCR has also been called RNA-PCR (Kawasaki, 1991), RNA phenotyping (Rappolee et al., 1991) and message amplification phenotyping (Brenner et al., 1989).

A number of methods such as Northern blots, RT-PCR, RNase protection assays, *in situ* hybridization, dot blots and S1 nuclease assays have been developed to measure gene expression in different tissues and cells. Of these techniques, RT-PCR is the most sensitive and is therefore frequently used to determine gene expression in different tissues particularly when the mRNA molecules of interest are present only in limited amounts either because few cells in the tissue express the sequence or because the transcript of interest is present in low copy number. In our studies, we have used RT-PCR for the detection of adrenergic receptor and neurokinin receptor gene expression in neurons of the central nervous system, sympathetic nervous system (Vidovic et al., 1994; Vidovic & Hill, 1997) and in some sympathetic peripheral targets such as the iris (Vidovic & Hill, 1995; Hill et al., 1996) and different arteries (Phillips et al., 1997). In these studies, the mRNA expression of different receptor subtypes was examined in the tissues of adult rats and compared to the expression pattern during development. The amount of tissue available per animal at early developmental stages is extremely small and these receptors may also be expressed at very low levels during development. For these reasons a very sensitive technique such as RT-PCR was used.

ISOLATION OF INTACT RNA

The key to successful RT-PCR is the isolation of intact RNA, free from genomic DNA, particularly when the mRNA of interest is the transcript of an intronless gene or when the genomic structure of a specific mRNA is unknown (Grillo & Margolis, 1990) or in cases where pseudogenes may be present. There are many methods for RNA extraction currently available in kit form. One widely used method developed by Chomezynski and Sacchi (1987), involves extraction of the guanidinium thiocyanate homogenate with phenol-chloroform at low pH. The Acid Guanidinium Thiocyanate-Phenol-Chloroform (AGPC) method is very rapid, thus suitable for isolating RNA from multiple tissues. However, some studies have reported genomic DNA contamination in RNA prepared by the AGPC method (Siebert & Chenchik, 1993). Attempts have therefore been made to either reduce the amount of DNA contamination by modification of the AGPC method (Siebert & Chenchik, 1993) or to eliminate the DNA by DNase treatment prior to RT-PCR (Grillo & Margolis, 1990). We have used 'RNAzol B' to isolate total RNA from different tissues for use in RT-PCR. 'RNAzol B' is one of several products developed by Cinna/Biotecx as a single reagent isolation method based on the AGPC method. This is a rapid and effective method for isolation of high yields of RNA from multiple samples and in situations where starting material may be limited. However, some genomic DNA contamination was detected when total RNA was isolated from some tissues, requiring treatment with DNase prior to RT-PCR (Phillips et al., 1997).

REVERSE TRANSCRIPTION OF RNA INTO cDNA

RNA sequences are amplified by the use of combined cDNA and PCR methods. A cDNA copy of an mRNA molecule is synthesized from preparations of either total RNA or mRNA (purified fraction of the total RNA) in a reaction that contains the enzyme reverse transcriptase (RNA dependent DNA polymerase). The two most commonly used enzymes for first strand cDNA synthesis are the avian myeloblastosis virus (AMV) or Moloney murine leukemia virus (MMLV or MuLV) reverse transcriptases. For RT enzymes to start reverse transcribing they require a small oligonucleotide such as either the oligo $(dT)_{12\text{-}18}$ or random primer (oligonucleotides of random sequences) to prime the reaction. Oligo (dT) anneals to the poly (A) tail at the 3′ terminus of mammalian mRNAs (Figure 30.1) and random primers, whose sequence diversity is so large that at least some primers will bind to the mRNA template, and prime the synthesis of the first strand of cDNA (Sambrook et al., 1989). RT reaction volume and the amount of RNA used varies depending on what protocol is used. In our laboratory we use 3–5 μg of total RNA in a reaction of 50 μl. One to 2 μl of the RT reaction is subsequently used in the PCR reaction.

Using RNA as a template can be problematic when reverse transcriptase is unable to synthesize cDNA from target RNA containing regions high in G+C or exhibiting complex secondary structure. Under these circumstances, the RNA must be denatured for efficient RT. Careful consideration should therefore be given to the type of enzyme system used for cDNA synthesis. The maximum recommended reaction temperature for MuLV is 42°C while the AMV enzyme is active at 48°C in optimized buffer conditions, thus eliminating some problems associated with complex secondary structures (Miller & Storts, 1995). In addition, many laboratories are characterising the RT activity of many thermostable DNA polymerases. Using thermostable DNA polymerases as reverse transcriptases would most likely eliminate the majority of problems associated with complex secondary structures.

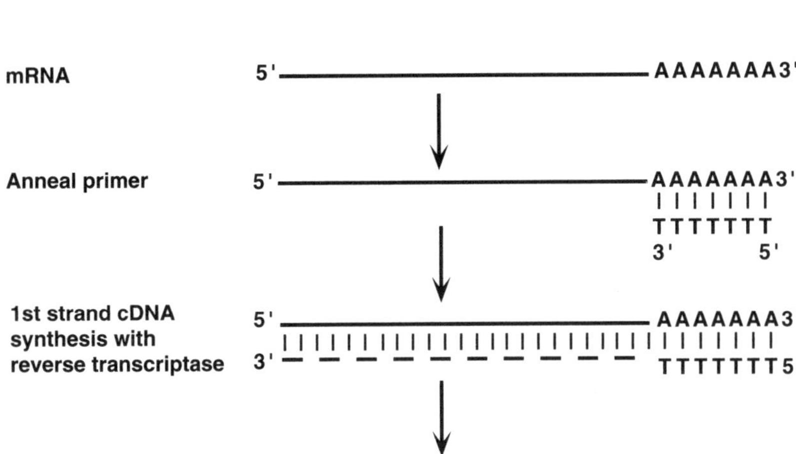

Figure 30.1 First strand cDNA synthesis. A cDNA copy of an mRNA molecule is synthesized by the enzyme reverse transcriptase. As illustrated, the reverse transcriptase requires a primer to start the synthesis. In this example, oligo (dT) anneals to the poly (A) tail at the 3′ terminus of mammalian mRNAs.

POLYMERASE CHAIN REACTION

Although, a number of one-step, one-buffer RT-PCR methods have been described (Miller & Storts, 1995) most RT-PCR protocols recommend a first strand cDNA synthesis reaction (as described above) followed by inactivation of the enzyme in the RT reaction and dilution of the RT mixture to eliminate inhibitory effects on the PCR process. The PCR part of the RT-PCR begins with the synthesis of the second strand of cDNA followed by amplification of specific regions of cDNA. Second strand cDNA synthesis and subsequent PCR amplification is achieved with purified, thermostable DNA polymerases. One most frequently used is *Thermus aquaticus* (*Taq*) DNA polymerase. The region for amplification is defined by two specific synthetic oligonucleotide primers, one primer complementary to each of the two cDNA or DNA strands. During the amplification process the two primers play a dual role: as a screening probe and amplification vector (Ruano et al., 1991). The principle of the PCR technique is illustrated in Figure 30.2. Each cycle of PCR begins with a brief heat denaturation (high temperature, 94°C) of the template DNA or cDNA and any previously synthesized product. A subsequent cooling of the template (temperature range 50–70°C) in the presence of a large excess of the two DNA primers allows their annealing to specific regions of nucleic acids. Once the annealing takes place, DNA polymerase anchors itself to the primer-template complex, deducts the four deoxyribonucleoside triphosphates from the

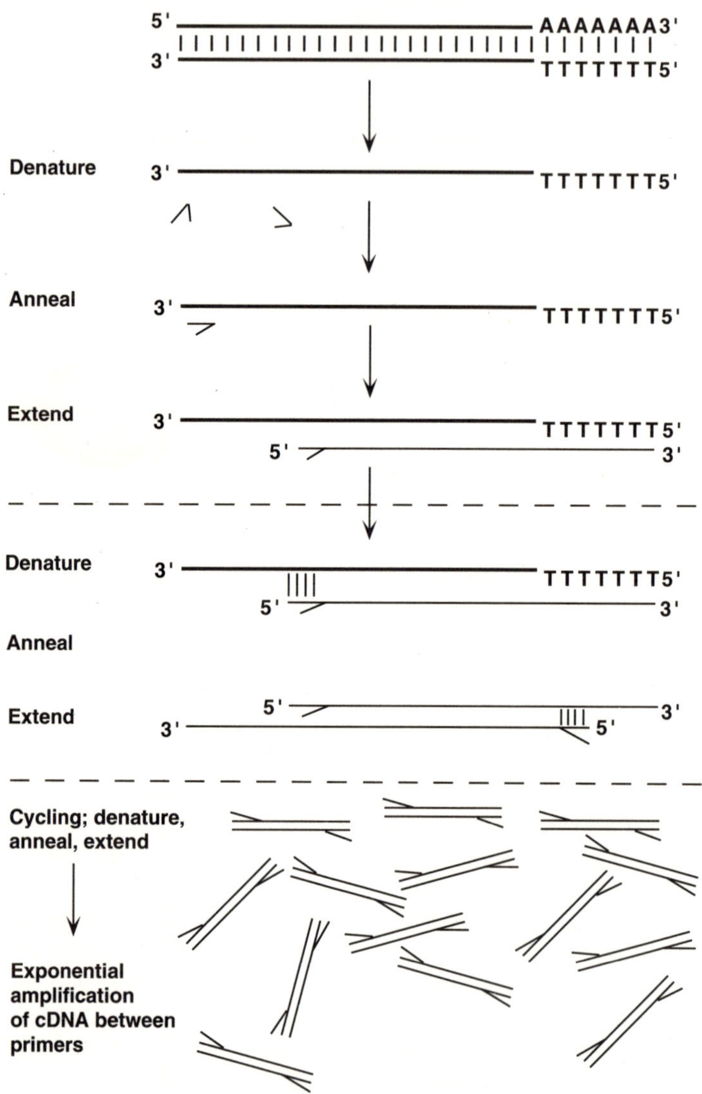

Figure 30.2 The polymerase chain reaction. Second strand cDNA synthesis and subsequent PCR amplification is produced by thermostable DNA polymerases. The RNA/cDNA hybrid molecule is heated to denature the complementary strands. A subsequent cooling of the template in the presence of a large excess of the two DNA primers allows their annealing to specific regions and the DNA polymerase extends from the primer along the template strand. After about 20–30 cycles of amplification a large amount of a single cDNA fragment, defined by the two primers, is generated.

medium and extends along the template strand. The extension temperature is set near the optimal for polymerase extension. For example, at 72°C for *Taq* DNA polymerase.

A variety of commercial PCR machines (or thermocyclers) are available. In our laboratory we use a capillary tube thermal cycler. Standard PCR reaction can be performed in a volume of 10–100 μl aliquots under standard conditions, usually provided by special buffer, supplied with DNA polymerase. For effective amplification, 20 to 30 cycles of reaction are required. Each cycle doubles the amount of DNA or cDNA synthesized in the previous cycle and this results in exponential increase in the target cDNA sequences. The amplification products are separated in 2% agarose gel stained with ethidium bromide and visualized with UV.

QUANTITATIVE RT-PCR METHODS

The RT-PCR method is a highly sensitive tool in the study of gene expression at the RNA level in mammalian tissues (McKinney & Robbins, 1992). However, obtaining quantitative information with this technique can be difficult. This is largely due to the nature of the two sequential enzymatic steps of RT-PCR. However, a number of quantitative RT-PCR methods have been described in the literature and are frequently used to determine the relative level of abundance or the absolute number of mRNA molecules of a particular mRNA in a sample. These include quantitative PCR with or without the use of internal standards, and competitive PCR. Some form of internal standards are used in quantitative PCR to control variations in amplification efficiency. The internal standard can be an endogenous sequence or gene transcript that is normally present in the sample, or an exogenous fragment added to the sample.

Determining relative levels of mRNAs using endogenous sequences, is done by co-amplifying a reference mRNA such as that of a 'housekeeping' gene (e.g. cyclophilin) with the target mRNA using two sets of specific primers. The amounts of PCR products generated by each set of primers are analyzed and compared. The abundance of the target gene transcript in a given sample can be expressed as a ratio of the abundance of the reference mRNA in the same sample (Burnet, 1994). Similarly, the exogenous sequences are co-amplified with target sequences using two sets of specific primers. The exogenous standard can either be a synthetic RNA added to the RT reaction or a DNA added to the PCR. Exogenous sequences are also used as templates in competitive PCR and share the same primer annealing sequences as the target nucleic acids and thus compete for the same primers and for amplification. Competitive PCR is the method of choice when the information on the absolute amount of target mRNA in a specific tissue is desired. Detailed description and discussion of these techniques is beyond the scope of this chapter and the reader is referred to Siebert (1993).

Although, most investigators use internal standards to control for variables in PCR and to obtain quantitative data on mRNA levels in a sample tissue, it is possible to do quantitative PCR without the use of internal standards. This method uses mathematical models and linear regression analysis to quantitate RNA levels (see Siebert, 1993, for detailed description and discussion of this approach).

REFERENCES AND FURTHER READING

Brenner, C. A., Tam, A. W., Nelson, P. A., Engleman, E. G., Suzuki, N., Fry, K. E. and Larrick, J. W. (1989) Message amplification phenotyping (MAPPing): A technique to simultaneously measure multiple mRNAs from small numbers of cells. *BioTechniques* **7**, 1096–1103.

Burnet, P. W. J., Eastwood, S. L. and Harrison, P. J. (1994) Detection and quantitation of 5-HT1A and 5-HT2A receptor mRNAs in human hippocampus using a reverse transcriptase-polymerase chain reaction (RT-PCR) technique and their correlation with binding site densities and age. *Neurosci. Lett.* **178**, 85–89.

Chomeczynski, P. and Sacchi, N. (1987) Single step method of RNA isolation by acid guanidinium thiocyanate-phenol-chloroform extraction. *Analytical Biochem.* **162**, 56–159.

Grillo, M. and Margolis, F. L. (1990) Use of reverse transcriptase polymerase chain reaction to monitor expression of intronless genes. *BioTechniques* **9**, 262–268.

Hill, C. E., Gould, D. J., Strigas, J., Burcher, E. and Vidovic, M. (1996) Sensory nerves play a role in the function of the arterioles, but not the dilator muscle, of the rat iris. *J. Auton. Nerv. System* **58**, 89–100.

Kawasaki, E. S. (1991) *PCR Protocols: A Guide to Methods and Applications*. M. A. Innis, D. H. Gelfand, J. J. Sninisky and T. J. White (eds). San Diego: Academic Press.

McKinney, M. and Robbins, M. (1992) Chronic atropine administration upregulates rat cortical muscarinic m1 receptor mRNA molecules: assessment with the RT/PCR. *Molecular Brain Res.* **12**, 29–45.

Miller, K. and Storts, D. R. (1995) PCR Access: A sensitive single tube, two-enzyme system for RT-PCR. *Promega Notes* **53**, 2–5.

Phillips, J. K., Vidovic, M. and Hill, C. E. (1997) Variation of alpha adrenergic, neurokinin and muscarinic receptors amongst four arteries of the rat. *J. Auton. Nerv. System* **62**, 85–93.

Rappolee, D. A., Mark, D., Banda, M. J. and Werb, Z. (1991) Wound macrophages express TGF-a and other growth factors *in vivo*: Analysis by mRNA phenotyping. *Science* **241**, 708–712.

Ruano, G., Brash, D. E. and Kidd, K. K. (1991) PCR: The first few cycles. *Amplifications* **7**,1–4.

Sambrook, J., Fritsch, E. F. and Maniatis, T. (1989) *Molecular Cloning. A Laboratory Manual*. New York: Cold Spring Harbor Laboratory Press.

Siebert, P. D. (1993) *Quantitative RT-PCR: Methods and applications*. Clontech Laboratories, Inc. **3**, 3–54

Siebert, P. D. and Chenchik, A. (1993) Modified acid guanidinium thiocyanate-phenol-chloroform RNA extraction method which greatly reduces DNA contamination. *Nucleic Acids Res.* **21**, 2019–2020.

Vidovic, M., Cohen, D. and Hill, C. E. (1994) Identification of alpha-2 adrenergic receptor gene expression in sympathetic neurons using polymerase chain reaction and *in situ* hybridization. *Molecular Brain Res.* **22**, 49–56.

Vidovic, M. and Hill, C. E. (1995) Alpha adrenoceptor gene expression in the rat iris during development and maturity. *Developmental Brain Res.* **89**, 309–313.

Vidovic, M. and Hill C. E. (1997) Transient expression of α-1B adrenoceptor mRNA in the rat superior cervical ganglion during postnatal development. *Neuroscience.* **77**, 841–848.

CHAPTER 31

DIFFERENTIAL DISPLAY OF GENE EXPRESSION

F. M. Freeman
Division of Biochemistry and Molecular Biology, John Curtin School of Medical Research, Australian National University

INTRODUCTION

This method is a combination of molecular biology techniques and is more specifically referred to as differential display reverse transcription—polymerase chain reaction (DDRT-PCR). It is used to identify genes whose expression is altered in response to a stimulus. The differential display method is a PCR-based method (see Chapter 30), and thus has the potential to identify very low levels of gene expression and hence rare species of mRNA. Performing differential display at a range of time points can help to elucidate a time course for the induction/repression of specific genes, and hence determine a biological cascade of events following stimulation.

REVERSE TRANSCRIPTION—POLYMERASE CHAIN REACTION (RT-PCR)

This method has been discussed in more detail in Chapter 30. Basically, messenger RNA (mRNA) is reverse transcribed, that is a strand of complementary DNA (cDNA) is synthesized from an mRNA molecule using the enzyme reverse transcriptase. This enzyme can only bind to double stranded RNA (or DNA), hence primers are employed. Primers are short strands of DNA (oligonucleotides) that hybridize with one end of a DNA (or RNA) molecule. The primer is said to be complementary to the DNA (or RNA) sequence to which it anneals (binds), that is, an adenine (A) (or uracil (U) in RNA) will only associate with thymine (T) and guanine (G) will only associate with cytosine (C). Usually they are designed based on the known sequence of a DNA molecule so that they are one hundred per cent complementary to their target DNA (or RNA) region. In differential display this is not possible as the candidate genes are not known.

Stimulated and control tissues are homogenized and their total RNA (messenger, ribosomal and transfer) extracted. Only mRNA is transcribed into proteins; it differs from the other species of RNA in that nearly all mRNAs have a multiple sequence of adenine at the 3′ terminal of its molecule, this is referred to as a poly A tail. Two per cent of the total RNA is mRNA, so some workers prefer to make a preparation of pure mRNA, but since the primers we employ target RNA with a poly A tail and hence only mRNA, this method of further purification can be omitted. Specifically, an RNA sample is reverse transcribed into cDNA (complementary DNA) using four sets of degenerate anchored oligo(dT) primers (T_{12}MN). M can be G, A or C and N is G, A, T or C. Each primer set is dictated by the 3′ base (N), with degeneracy in the penultimate (M) position. For example, the primer set where N = C consists of:

5′-TTTTTTTTTTTTGC-3′
5′-TTTTTTTTTTTTAC-3′
5′-TTTTTTTTTTTTCC-3′

The initial RNA sample thus gives rise to four cDNA samples consisting of either T_{12}MA, T_{12}MC, T_{12}MG or T_{12}MT 3′ end tails. To amplify the cDNAs by PCR a set of arbitrary primers, as well as the original anchor primers, are employed. The arbitrary primer binds to the 5′ end of the cDNA molecule and can be of any length but usually consist of between 10–12 bases. Bauer et al. (1993) proposed the use of 26 different arbitrary decamer primers to identify the entire population of mRNA, approximately 12 000–15 000 distinct RNA species per eukaryotic cell, from a single sample. Since the primers used here are shorter than for conventional PCR, the temperature for DDRT-PCR is lower than usual, at 42°C. In addition to the newly synthesized cDNA, anchor and arbitrary primers the radio label ^{32}P-ATP is added. This is incorporated into the newly synthesized strand of cDNA and is used to identify it. The resulting samples are then loaded onto a polyacrylamide gel and electrophoresized. This separates the DNA based on size. The gel is dried onto filter paper and apposed to X-ray film, giving rise to an autoradiogram of the gel. The bands correspond to radio labelled cDNA, reverse transcribed from the original mRNA. By comparing the bands in stimulated (S) and control (C) tissue, one can identify genes that are differentially expressed (Figure 31.1).

VERIFICATION OF CANDIDATES

Many of the cDNA bands that appear to be differentially expressed turn out to be false positives (between 50 and 70%), that is the band is not differentially expressed or the sample has been contaminated prior to PCR by a DNA molecule not present in the original sample. Hence one needs to determine whether the candidate is legitimate or not. Differentially expressed bands are cut out from the gel, their DNA eluted and further amplified by PCR, using the anchor and arbitrary primers used in the initial isolation. This DNA can then be used to probe, and hence determine whether it was differentially expressed in the original sample. Firstly the candidate/probe has to be labelled for easy identification. There are many methods of labelling. One of the most straightforward methods is random labelling, using the Klenow fragment of DNA polymerase 1. A radio labelled nucleotide (^{32}P – dATP) randomly replaces an unlabelled nucleotide in the DNA molecule. Northern blot analysis (see Chapter 26), where the original RNA samples (stimulated and control) is electrophoresized, transferred to a membrane and probed with the now radiolabelled candidate, thus determining whether the candidate is differentially expressed due to the stimulation. Occasionally Northern blotting gives rise to no bands, this could be due to the method being insensitive to rare mRNA species. In that case the more sensitive method of quantification, RT-PCR, is employed. Here one must sequence the candidate so that more specific, longer primers can be designed for its PCR. The original primer pair (anchor and arbitrary) cannot be used as they will amplify all the bands they recognize from the original sample. In PCR, DNA amplification is only exponential when enough template is present, or reagents are in excess. Hence samples are

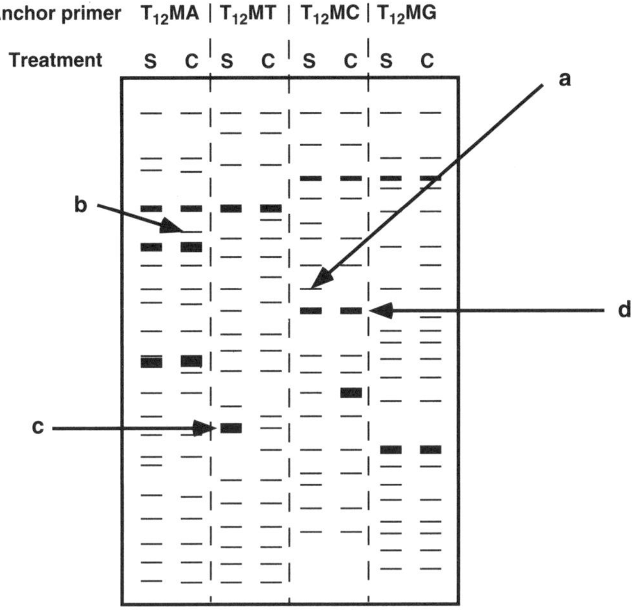

Figure 31.1 Schematic representation of an autoradiogram showing cDNA bands that are differentially displayed using one arbitrary primer and the four anchor primers. Examples of genes whose expression is either (a) induced, (b) repressed or (c) ehanced due to the stimulus. Some genes (d) are highly expressed in both stimulated (S) and control (C) samples.

taken after different cycles to ensure that comparisons between samples are made during this linear phase.

IDENTIFICATION OF CANDIDATE

A differentially expressed candidate can be identified by the sequence of its bases. The candidate has to be cloned prior to sequencing. This allows the addition of known sequences of DNA at either end of the candidate, which are essential for sequencing. Cloning involves the insertion of the candidate, using restriction enzymes, into a piece of circular DNA, referred to as a plasmid vector. The vector is taken up by a bacterial cell, such as *E. coli*. The bacteria treats the plasmid as its own DNA, so that when it replicates it also replicates the plasmid.

After a few rounds of replication the bacteria is chemically lysed, its DNA removed and plasmid isolated. Using restriction enzymes, the candidate is excised from its plasmid. Sequencing of the DNA is achieved using the chain-terminator method (Sanger et al., 1977). Basically, the candidate is amplified using DNA polymerase as usual, except an extra reagent, 2′3′-dideoxynucleoside triphosphate, is added. Once this analog is

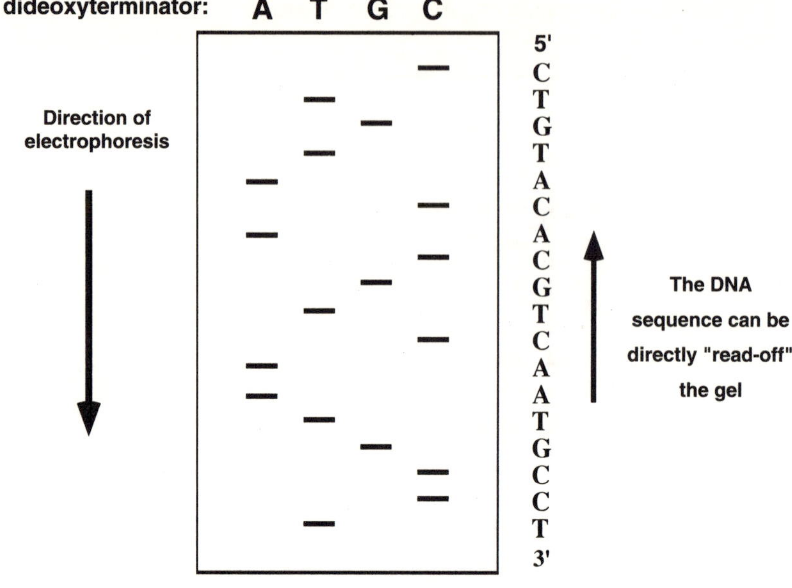

Figure 31.2 DNA sequencing using dideoxynucleoside triphosphates as chain terminators. Band position, and hence base order can be directly read from the gel.

incorporated into at the growing point of the DNA chain it can add no more dNTPs. Hence synthesis of this particular DNA molecule is terminated. DNA synthesis is carried out in the presence of labelled dNTPs and either of the 4 dideoxynucleoside triphosphate analogues. Therefore, in each reaction there is a population of particle synthesized DNA, each having a common 5′ end but each varying in length to a base-specific 3′ end. These samples can then be electrophoresized side by side. The sequence can be read off the resultant auto radiogram (Figure 31.2).

Alternatively, to radiolabeling, fluorescent dyes can be used to label the DNA and the samples electrophoresized. A fluorescence detector connected to a computer then notes the fluorescent bands as they 'run off' the bottom of the gel, thereby directly sequencing the candidate. The nucleotide sequence from either technique can then be compared for homology with known sequences of genes that are listed in the GeneBank database, located at the National Institute of Health in America. Completely novel gene sequences can be further studied to (a) determine their complete sequence for depositing into the genebank database and (b) identify their role in the cell.

DIFFERENTIAL EXPRESSION OF KNOWN GENES

Obviously, if one knows the identity of the genes that will be differential expressed, then more specific primers, than the anchor or arbitrary primers, can be designed and employed. This has the advantage of increasing the temperature for PCR and decreasing the degree of 'miss-matching' and hence the number of false positive bands.

ADVANTAGES AND DISADVANTAGES

Advantages

A major advantage of differential display is that less than 5 mg of total RNA is needed to amplify all the mRNAs in a given cell type. In contrast, other methods of studying gene expression require 20–100 times more total RNA. Also, differential display is a highly sensitive method for the identification of novel genes expressed in response to a stimuli. It can identify rare mRNA species. In fact, some scientists employ the technique of *in vivo* DDRT-PCR, where gene expression from single neurons can be compared. This eradicates any dilution factor problems where rare mRNA are diluted out with genes from surrounding unstimulated tissue. It is approximately the same as *in vitro* DDRT-PCR except that advantage is taken of the enzymes and reagents present within a single living cell. This method usually combines the electrophysiological recording technique of patch-clamping, where the clamping pipette can be use to introduce reagents and/or remove the contents of the cell during or after DDRT-PCR. This has the advantage of being able to study the activity of a single neuron and relate its behavior to its gene expression.

Disadvantages

The major disadvantage of this method is the high number of 'false positive' candidates; between 50 and 70% or even higher in some cases. This is exacerbated by the number of technically demanding steps in the procedure. A few examples of these disadvantages are:

1. PCR is a very sensitive method that is capable of amplifying a single molecule of DNA and any contamination of samples results in false positives. In an attempt to overcome this, PCR is conducted in an sterile environment and at least duplicate samples of stimulated and control tissue run in parallel.
2. The low temperature of PCR results in a high proportion of primer binding to miss-matched sequences.
3. In the initial screening for differentially expressed candidates, when a band is excised from the gel it may not contain only one species of DNA. In fact it could contain DNA that is not differentially expressed. Hence during candidate verification one will not know if only one species of DNA is being labelled and hence differentially expressed. Some DNAs, which may not be the one in excess within the band, are preferentially taken up by the plasmid. To try and overcome this problem cloning has to be repeated several times and the sequenced candidate's legitimacy re-evaluated.
4. Whilst DDRT-PCR is a useful and very sensitive tool, the transcription of a gene does not guarantee its translation into protein, and hence involvement in the biological

cascade. Thus further protein studies have to be performed to determine whether the gene plays a role in the cascade of events that follow a stimulus.

RECENT PAPER

Recently Inokuchi et al. (1996) identified a gene, activin βA, that is induced in response to exposure to the chemical kainate. Kainate agonizes a sub-type of glutamate receptors. Application of kainate to brain tissue causes generalized seizures and is an experimental model for the study of epilepsy. Basically a rat received an intraperitoneal injection of kainate and was allowed to convulse for 3 hours. A region of the rat's brain, the hippocampus, was dissected out and differential display performed on kainate exposed and naive rat tissue. Activin βA was identified as being differentially expressed; this was confirmed using Northern blot analysis. This molecule had not previously been identified as being involved in the cascade of biological events that follow exposure to kainate. In addition, Inokuchi et al. showed that this molecule can also be induced when hippocampal neurons are electrically stimulated, and that its induction is mediated via activation of glutamate receptors. In non-neuronal cells it has been shown that activin βA is a member of the transforming growth factor-β (TGF-β) superfamily. Whether it plays a similar role in brain hippocampal tissue remains to be determined.

REFERENCES AND FURTHER READING

Bauer, D., Muller, H., Reich, J., Riedel, H., Ahrenkiel, V., Warthoe, P. and Strauss M. (1993) Identification of differentially expressed mRNA species by an improved display technique (DDRT-PCR). *Nucleic Acid Res.* **21**, 4272–4280.

Guimaraes, M. G., Lee, F., Zlotnik, A. and McClanahan, T. (1995) Differential display by PCR: novel findings and applications. *Nucleic Acid Res.* **23**, 1832–1833.

Inokuchi, K., Kato, A., Hiraia, K., Hishinuma, F., Inoue, M. and Ozawa, F. (1996) Increase in activin βA mRNA in rat hippocampus during long-term potentiation. *FEBS Letters* **382**, 48–52.

Liang, P., Averboukh, L. and Pardee, A. B. (1993) Distribution and cloning of eukaryotic mRNAs by means of differential display: refinements and optimization. *Nucleic Acid Res.* **21**, 3269–3275.

Liang, P. and Pardee, A. B. (1992) Differential display of eukaryotic messenger RNA by means of the polymerase chain reaction. *Science* **257**, 967–971.

Livesey, F. J. and Hunt, S. P. (1996) Identifying changes in gene expression in the nervous system: mRNA differential display. *Trends Neurosci.* **19**, 84–88.

Sanger F., Nicklen S. and Coulson A. R. (1977) DNA sequencing with chain terminating inhibitors. *Proc. Natl. Acad. Sci.* **74**, 5463–5467.

Watson, J. B. and Margulies, J. E. (1993) Differential cDNA screeening strategies to identify novel stage-specific proteins in the developing mammalian brain. *Developmental Neurosci.* **15**, 77–86.

CHAPTER 32

INHIBITION OF GENE EXPRESSION WITH ANTISENSE OLIGONULCLEOTIDES

P. M. Pilowsky
Department of Neurosurgery, Royal North Shore Hospital, Sydney

INTRODUCTION

The idea of using ANTISENSE oligonucleotides to prevent translation of messenger ribonucleic acid (mRNA) is strikingly simple. However, as with any technique there are difficulties that must be considered. The basic principle is illustrated in Figure 32.1. Essentially, if the mRNA sequence transcribed and processed from a gene is known, then a short sequence of DNA can be constructed that will hybridize to the mRNA. This hybridization, or binding, prevents the mRNA from attaching to ribosomes and may also enhance mRNA degradation by intracellular RNase's. The end result is a decrease in gene expression.

The term ANTISENSE, in relation to strands of the nucleotide bases (thymine, cytosine, guanine and adenine), refers to any sequence of these bases that is complementary to the corresponding SENSE sequence of a gene. For example, the genome itself is composed of a SENSE and an ANTISENSE strand. Information is *transcribed* from the ANTISENSE strand resulting in a strand of RNA that contains the same information as the SENSE strand of DNA. This RNA is known as heterologous nuclear RNA (hnRNA). Removal of introns (non-expressed portions) and splicing of exons (expressed portions) results in mRNA. The sequence in the mRNA is then translated into protein, or *gene product*, after binding to ribosomes in the cytoplasm. Sometimes, different species of mRNA can be formed from a single gene by differential splicing of hnRNA. Oligonucleotides are short strands of DNA. These include oligonucleotides that are used to detect mRNA in *in situ* hybridization experiments. Here we are only concerned with oligonucleotides used for attenuation of gene expression *in vivo*. Many experiments using ANTISENSE have now been conducted both *in vivo* and *in vitro*, in organisms ranging from plants to mammals and with an enormously diverse range of behaviors. It should be noted that the success rate has been equally diverse—more on this later. It is also possible that ANTISENSE sequences may be a part of normal cellular physiology in some micro-organisms.

The use of ANTISENSE to attenuate gene expression began in the mid-1980s among immunologists and microbiologists (e.g. Loke et al., 1989). The technique continues to find wide application in these fields. Amongst neuroscientists, initial studies were confined to cell culture studies in the late-1980s and finally, *in vivo* studies by the early-1990s. Several excellent reviews are now available covering recent developments in ANTISENSE technology and use (Albert & Morris, 1994; Chiasson et al., 1994; Pilowsky et al., 1994; Wahlestedt, 1994; Agrawal et al., 1995).

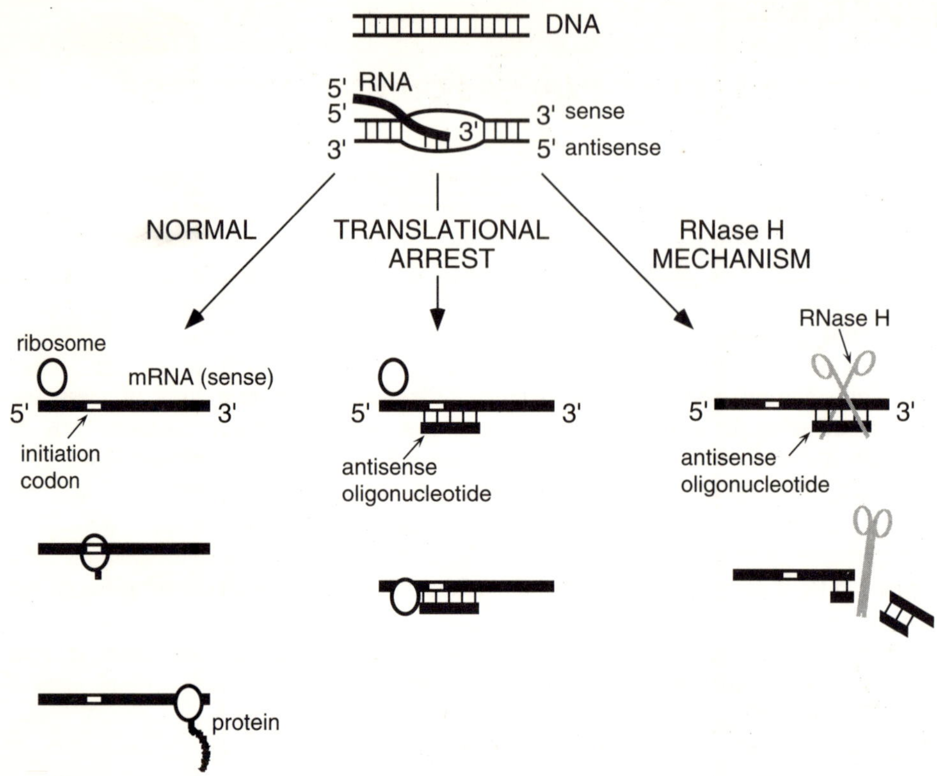

Figure 32.1 A representation of the process of normal gene expression and two mechanisms by which ANTISENSE oligonucleotides might attenuate translation of protein from mRNA. DNA has two strands: a SENSE strand and an ANTISENSE strand. The two strands are complementary. Messenger RNA is copied from the ANTISENSE strand, so that its sequence is the same as the corresponding SENSE sequence of the DNA (except for conversion of base thymine in DNA to base uracil in RNA). ANTISENSE oligonucleotides have the same sequence as a portion of the ANTISENSE strand of DNA, so that they bind to the messenger RNA preventing its translation by ribosomes (TRANSLATIONAL ARREST). Alternatively, hybrids of ANTISENSE and mRNA might be more rapidly degraded in the cytoplasm by the enzyme RNase H (RNase H MECHANISM). (Reprinted from Pilowsky et al., 1994 with permission from Blackwell Science Pty Ltd.)

DESIGN OF ANTISENSE OLIGONUCLEOTIDES

ANTISENSE oligonucleotides are normally made so that they will span the initiation codon of the targeted mRNA. Several studies have shown that this is the most effective strategy. Sequences directed at other sites tend to be less effective.

A number of other factors need to be considered when preparing ANTISENSE oligonucleotides. These include length, base composition and prevention of degradation.

Longer oligonucleotides may hybridize more effectively to their target. On the other hand if the length is too great, then penetration into cells may be a problem. Unlike *in situ* hybridization where lengths of around 40 to 50 for cDNA probes are common. ANTISENSE oligonucleotides for inhibition of gene expression are commonly 15–18 bases long. Degradation of oligonucleotides is a well recognized occurrence, although its importance is debatable. In plasma the half-life of unmodified cDNA is quite short, but experiments using microinjections of unmodified cDNA into brain have been quite successful.

The specificity of ANTISENSE oligonucleotides can depends on the sequence chosen. Many genes belong to families and ANTISENSE oligonucleotides can be designed to include many members of the family, or to select for only one. A similar strategy is used in immunohistochemistry where it may be useful to have a highly selective antibody to pick out one unique epitope or a less selective (but still specific) antibody to pick out a unique epitope that is common to several classes of protein (Albert & Morris, 1994). Pharmacologists also use this approach with combinations of selective and non-selective (but still specific) drugs to investigate receptor profiles.

Several recent reviews have canvassed issues such as doses, routes of administration, target genes, ANTISENSE design and delivery. The reader is referred to these for more detailed information. Furthermore, as a field in its infancy, attenuation of gene expression is rapidly evolving. ANTISENSE appeared as a keyword at the United States Society for Neuroscience meeting on 7 papers in 1991. This number grew to 62 in 1995 as other groups experimented with ANTISENSE attenuation of gene expression. It remains to be seen how this technique, which occupies an unusual niche between pharmacology, physiology and molecular biology, will evolve.

CONTROLS FOR ANTISENSE OLIGONUCLEOTIDES

As with any experiment, appropriate controls are critical. In ANTISENSE experiments a common control is to use an oligonucleotide with the same base composition but with one or two bases switched around in the sequence. This is known as a MISMATCH, and is probably the safest control available, since even a one base mismatch will cause a 500-fold decrease in binding efficiency which should be more than adequate to eliminate any specific effects of the ANTISENSE sequence. However, in a word of warning, Chiasson et al. (1994) noted that in their experiments with the oncogene *c-fos*, even a 2–3 base substitution could still confer antisense activity on an oligonucleotide. As may be expected, a vehicle control is essential to monitor possible non-specific effects of ANTISENSE oligonucleotides. Two alternatives to MISMATCH oligonucleotides are SENSE, which is the complementary sequence to ANTISENSE, and NON-, JUMBLED-, or SCRAMBLED ANTISENSE which is the ANTISENSE sequence in a completely jumbled sequence. Some investigators like to use combinations, or all, of these controls. Such findings suggest that caution is warranted and that several oligonucleotides and other controls may be required to demonstrate convincingly that an effect is due to a specific attenuation of gene expression.

OBTAINING AND STORING OLIGONUCLEOTIDES

Oligonucleotides can be obtained from several commercial sources and are stored in water at –20 to –80°C. They are extremely physically stable—much more so than proteins—but must be protected from nucleases which are often found in serum or blood.

EFFECTIVENESS OF ANTISENSE OLIGONUCLEOTIDES

There are several ways to test for the effectiveness of the ANTISENSE oligonucleotides in an experimental situation. If an antisera is available, then a reduction in detectable protein can be assayed. This can be done immunohistochemically or by Western blotting. Unfortunately, the extent of reduction in cellular protein concentration may not correlate precisely with effectiveness in a physiological test. For example a fifty percent reduction in expression of an enzyme could result in no physiological effect if the enzyme has a high turnover rate and is not rate-limiting. On the other hand, reductions in the concentration of voltage gated ion channels may result in highly correlated reductions in responses when agents that affect such channels are applied. Thus, Wahlestedt et al. (1993) found that pretreatment of rats with ANTISENSE oligonucleotides to the *N*-methyl-d-aspartate (NMDA) receptor R1 subunit was approximately as effective at reducing the size of an experimental ischaemic infarction of the cortex of the brain as treatment with MK801, a selective inhibitor of the NMDA receptor.

Measurement of mRNA concentrations by Northern blotting may be a useful way to establish if there is a change in mRNA concentration but again may not help in determining physiological effectiveness. Perhaps the key question is the specificity of ANTISENSE oligonucleotides as pharmacological agents in a given experimental situation.

RECENT PAPER

ANTISENSE oligonucleotides are often useful in situations where no highly specific antagonist of a protein's action is available. This is well illustrated in a recent report by Berhow et al. (1996) ANTISENSE oligonucleotides to a brain kinase (ERK (Extracellular signal Regulated Kinase)) was used in order to investigate, amongst other things, the role of ERK in a neurotrophin signal transduction cascade. Infusion of ANTISENSE oligonucleotides to ERK1 into the ventral tegmental area resulted in a reversible fall in ERK immunoreactivity of around 35–40%. No effect was seen with vehicle, SENSE, or SCRAMBLED ANTISENSE, infusions. Despite this apparently modest reduction in immunoreactivity, ANTISENSE oligonucleotides to ERK completely blocked the ability of morphine to increase the activity of ERK. The authors attribute this finding to an increase in the fraction of ERK molecules in the phosphorylated state.

SUMMARY

ANTISENSE is a new technique that aims to reduce the expression of specific genes. This aim is achievable to varying degrees in a variety of circumstances, but care is required in the choice of ANTISENSE and controls. ANTISENSE is particularly useful for studying the physiology of a gene product when molecular biologists have established the gene sequence, but where no specific pharmacology exists for interfering with the function of the gene product. Such gene products may include neurotransmitters, receptors, transporters, enzymes and intracellular signalling pathways.

REFERENCES AND FURTHER READING

Agrawal, S., Temsamani, J. Galbraith, W. and Tang. J. (1995) Pharmacokinetics of antisense oligonucleotides. *Clin. Pharmacokinetics* **28**, 7–16.

Albert, P. R. and Morris, S. J. (1994) Antisense knockouts: molecular scapels for the dissection of signal transduction. *Trends Pharmacol. Sci.* **15**, 250–254.

Berhow, M. T., Hiroi, N. and Nestler, E. J. (1996) Regulation of ERK (extracellular signal regulated kinase), part of the neurotrophin signal transduction cascade, in the rat mesolimbic dopamine system by chronic exposure to morphine or cocaine. *J. Neurosci.* **16**, 4707–4715.

Chiasson, B. J., Armstrong, J. N., Hooper, M. L.,Murphy, P. R. and Robertson, H. A. (1994) The application of antisense oligonucleotide technology to the brain: some pitfalls. *Cell. Molec. Neurobiol.* **14**, 507–521.

Loke, S. L., Stein, C. A., Zhang, X. H., Mori, K.,Nakanishi, M.,Subasinghe, C., Cohen, J. S. and Neckers, L. M. (1989) Characterization of oligonucleotide transport into living cells. *Proc Natl Acad. Sci.* **86**, 3474–3478.

Pilowsky, P. M., Suzuki, S. and Minson. B. (1994) Antisense oligonucleotides: A new tool in neuroscience. *Clin. Experimtl. Pharmacol. Physiol.* **21**, 935–944.

Wahlestedt, C. (1994) Antisense oligonucleotides strategies in neuropharmacology. *Trends Pharmacol. Sci.* **15**, 42–46.

Wahlestedt, C., Golanov, E., Yamamoto, S., Yee, F., Ericson, H., Yoo, H., Inturrisi, C. E. and Reis, D. J. (1993) Antisense oligodeoxynucleotides to NMDA-R1 receptor channel protect cortical neurons from excitotoxicity and reduce focal ischaemic infarctions. *Nature* **363**, 260–263.

CHAPTER 33

CREATION OF TRANSGENIC MICE

P. G. Noakes
Department of Physiology and Pharmacology, University of Queensland

INTRODUCTION

The use of transgenic animals in biological and biomedical research has become widespread in recent years. The applications are numerous, ranging from the generation of animal models for known human genetic disorders, to understanding the elements that control tissue specific expression of the genes in the developing animal. These sorts of applications are now being pursued by neuroscientists in their research efforts.

Transgenic animals

The animal of choice for transgenic work is the mouse. Mice are cheap, require little space, breed quickly, and their genetics have been well characterised. In recent years, other transgenic animals such as transgenic cows, pigs, chickens and rats have also been generated.

Types of transgenics

Basically there are two types of transgenics. The first type is generated by the introduction of a foreign gene into the developing animal's genome, and is what most people mean when they talk about transgenics. It is usually achieved by direct injection of a fragment of DNA containing a recombinant gene into the male pronucleus of a fertilised egg. The second type is where an animal's particular endogenous gene is mutated. This type of transgenic is commonly referred to as a gene targeted animal. In this type, mutations to the gene of interest can either render it useless, so that no protein is encoded (a knockout or null mutation), or create a mutated form of the protein where functional domains are altered (a hit and run; see Hasty & Bradley, 1992).

There are many fine reviews and books describing in detail the techniques and applications of transgenic animals; these are listed at the end. This article is only intended as an introduction to what is becoming a powerful method in neuroscience research. The following is a brief overview of gene targeted (knockout type) transgenic mice.

GENE TARGETED TRANSGENICS

Why generate a gene targeted animal?

The major reason is to assess the function of a molecule directly in the living animal. Other reasons include the desire to generate an animal model for a known genetic disease.

In other words, if a known genetic disorder is due to a mutated gene, then one can engineer a mouse carrying the same mutation. This mutant mouse could then become an animal model for the genetic disease upon which various strategies can be tried to alleviate the condition, either via drugs or by the reintroduction of a 'good' copy of the gene into the mouse (gene therapy).

However, before entering into the time consuming task of making a gene targeted mouse, it is well worth the effort to find out as much about the biology of the molecule that is being targeted, as this will largely determine how the knockout mouse is to be assessed. For example, knowing about a particular molecule's spatial and temporal expression during development could well determine when a potential phenotype becomes apparent. For neuroscientists, this is an important issue, as a molecule may be expressed and be essential for life, before the birth of the nervous system (gastrulation), and thus could hamper their investigations. However, there are new advances in gene targeting that can get around this potential problem, such as tissue and inducible specific knockouts (see below).

Homologous recombination

The ability to either inactivate or mutate a functional portion of the gene of interest relies on homologous recombination between the manipulated portion of the cloned gene and the endogenous gene from which this clone was first isolated (Figure 33.1). This is a rare event occurring at a frequency of between 10^{-2} to 10^{-7} of cells transfected.

Factors that affect homologous recombination (HR) in mammalian cells are: the strain of DNA used to manipulate must be the same as the DNA of the gene to be targeted (isogenic DNA); the length of homology of the manipulated DNA should be at least 5 kilo bases; and the cells to be transfected must be undergoing cell division, as HR occurs during the S phase of cell division (Capecchi, 1989).

METHOD

Stage 1: Pulling out a gDNA

Making a gene targeted mouse begins with isolating a genomic clone of the gene to be targeted.

A cloned portion of the gene (cDNA) is usually used as a probe to isolate recombinant lambda bacteriophage clones containing parts of the gene. Where the aim is to produce a non-expressing form of the gene (null mutation) the 5 prime end of the gene containing the first few exons is cloned. However, if the aim is to mutate a functional portion of the gene, then the part of the gene that spans exons encoding for such a domain should be isolated using the appropriate part of the cDNA.

As mentioned earlier, it is imperative that the mouse genomic library should be from the same mouse strain as the mouse cell that will eventually be transfected. This will ensure 100% homology sequence between the targeting construct and the gene to be targeted.

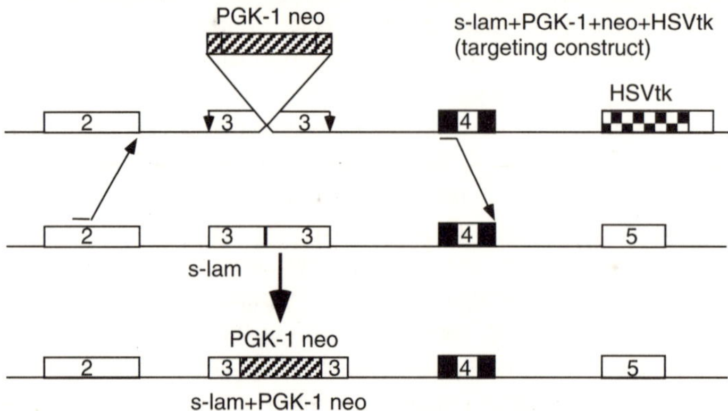

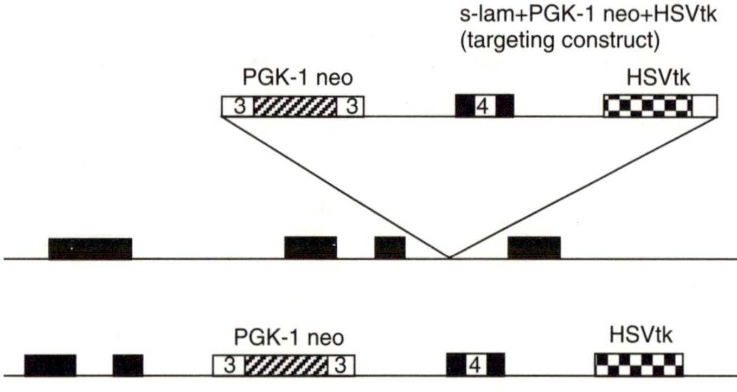

Figure 33.1 Schematic diagram showing homologous recombination (A) and random integration (B) events. A shows homologous recombination between the targeting construct made from cloned genomic sequences of the s-laminin gene (s-lam+PGK-1+neo+HSVtk), and the endogenous s-laminin gene (s-lam). In this event, the HSVtk gene (chequered box) is lost during the recombination event, due to non-homology. B shows random integration whereby the entire targeting construct including HSVtk is incorporated into virtually any part of the genome at random. The targeting construct shown is of the replacement type, where the mutation, PGK-1 neo (hatched box) is inserted into the third coding exon of a s-laminin gene clone. PGK-1 neo consists of the PGK-1 (phospoglycerate kinase-1) promoter which is used to drive high levels of expression of the bacterial gene Neomycin phosphotransferase (neo). The neo gene encodes for drug resistance to geneticin (G418), allowing for selection of clones of embryonic stem (ES) cells that have integrated the targeting construct (either targeted or random). At the end of the construct, the viral gene HSVtk encoding for drug sensitivity to ganciclovir is added. The loss of HSVtk allows for the selection of ES cell clones where targeted events have occurred, as the drug ganciclovir will kill off clones that have incorporated an active HSVtk gene (adapted from Mansour et al., 1988).

Stage 2: Making a targeting construct

Normally, the length of gDNA needed to make a targeting construct is about 5 to 6 kilobases long. This amount of gDNA is not easy to manipulate and takes the experience of a good molecular biologist to make a targeting construct.

The first requirement is that a detailed analysis of the gDNA clone be done. This is achieved by restriction mapping and sequencing of the gDNA clone. Once this is known it is possible to cut the gDNA at a specific site, usually in a coding exon, to then allow insertion of the mutation of choice. This manipulated piece of gDNA is referred to as a targeting construct. There are basically two types of targeting constructs that can be made, either a replacement type or an insertion type (see Hasty & Bradley (1992) and Hogan et al. (1994) for details). Only the replacement type will be dealt with here, as this is most commonly used.

The replacement type of targeting construct involves simply inserting the mutation into a coding exon within the gDNA clone with gDNA sequences flanking either side of the mutation. Usually one of the flanking sequences is shorter than the other for ease of identification of HR events (Figure 33.1).

When the targeting construct is introduced into embryonic stem cells it can become integrated into the chromosomes in one of two ways. Random integration is the most common, with the entirety of the targeting construct being incorporated into virtually any part of the genome in an unpredictable way. This is the kind of integration that is used when making transgenic mice by pronuclei injections. The second type of integration is homologous recombination (HR) in which the targeting construct DNA containing mutations to the gene of interest replaces part of the homologous endogenous gene (Figure 33.1). Homologous recombination is a much rarer event than random integration (see above).

In order to enrich for HR events over random integration events an extra gene, the Herpes simplex thymine kinase (HSVtk) gene, is added at the end of the targeting construct. The HSVtk gene encodes the thymine kinase, which is sensitive to the drugs ganciclovir (GANC) or 1-(2′-deoxy-2′-fluoro-β-D-arabinofuranosyl)-5-iodouracil (FIAU). Any cell that is expressing this protein can be killed off by these drugs. In random integration events, the HSVtk gene at the end of the targeting construct is incorporated and expressed. In contrast, during homologous recombination events the HSVtk gene is not integrated along with the rest of the gene due to non-homology. Thus this strategy is used to enrich for HR events over random integration events (Cappechi, 1989). In practice however, this type of selection (negative selection) only enriches for HR events by about 2 to 5 fold over random integration, due to the HSVtk gene being inactivated (not lost) during random integration. In an attempt to get around this problem, some investigators are now putting two HSVtk genes at each end of the construct to increase the enrichment of HR events over random events.

In recent years, these replacement vectors have had a couple of modifications that have improved the chances of making and studying knockout mice. First, deletions have been made in the gDNA sequences that flank the mutation. This drops the overall homology, but ensures that if the mutation is spliced out, or if there is read through the stop codons, then subsequent transcription of the gene is out of the translational reading frame and results in a non-functional protein. Second, investigators are now using a neo-gene that is fused to a reporter gene, usually *lac*Z. The advantage of this is that the cells that express the mutated gene (those that would express the normal version of the gene) can be identified *in vivo* because *lac*Z encodes for the enzyme β-galactosidase which can be histochemically detected in tissues. This is a good strategy if you do not know all the tissues and cell types that might express your gene, and aids in the analysis of the gene targeted animal, by allowing examination of cells and tissues that express the reporter gene for abnormalities, etc.

Stage 3: Embryonic stem cells

Once armed with a targeting construct, the aim is to transfect mouse cells then generate a mutant mouse. Cells appropriate for transfection are derived from the inner cell mass of a developing embryo and are termed embryonic stem (ES) cells (Evans and Kaufam, 1981; Martin 1981). Bradley, Robertson, and colleagues (Bradley et al., 1984; Robertson et al. 1986) showed that these cells could be grown in culture for varying periods of time, manipulated, and then be reintroduced back into a developing mouse embryo (blastocyst stage) to participate in the embryo's normal development and contribute to every tissue. In particular, to the sex cell progenitor (germ cell) population of cells giving rise to the gametes of the mouse.

For these ES cells to retain this unique characteristic of re-colonising all tissues such as the germ line population, they must be cared for in culture and preferably not go through too many passages in culture (splitting and the re-plating of cells) from the time they were first derived from an embryo. This is termed passage number of the ES cell.

Stage 4: Gene targeting in embryonic stem cells

To gene target ES cells, the method of choice is to transfect these cells with a linearized version of the targeting construct by electroporation. The efficiency of transfecting cells in this manner is about 1 per 1 000 cells for random integration. Electroporated cells are then plated out and screened for HR events (Figure 33.2).

Stage 5: Screening for homologous recombination events between the targeting construct and the gene that is to be targeted

The screening is based upon drug selection using geneticin (G418) and ganciclovir (or FIAU) over a period of ten days. During this time, ES cells that have been successfully transfected (both random and targeted) will have an active neo gene and thus grow in the presence of G418. The addition of GANC (or FIAU) will kill any ES cell that has incorporated an active HSVtk gene via random intergration. As mentioned previously, the HSVtk gene is frequently inactivated, leading to a number of GANC resistant ES cell clones that are not targeted (Figure 33.2).

The only sure way of screening for HR events in ES cells, is to examine the ES cell's gDNA at the intended site of integration. Many investigators use a two step approach. Firstly, a polymerase chain reaction (PCR) to amplify the targeted region of the genome

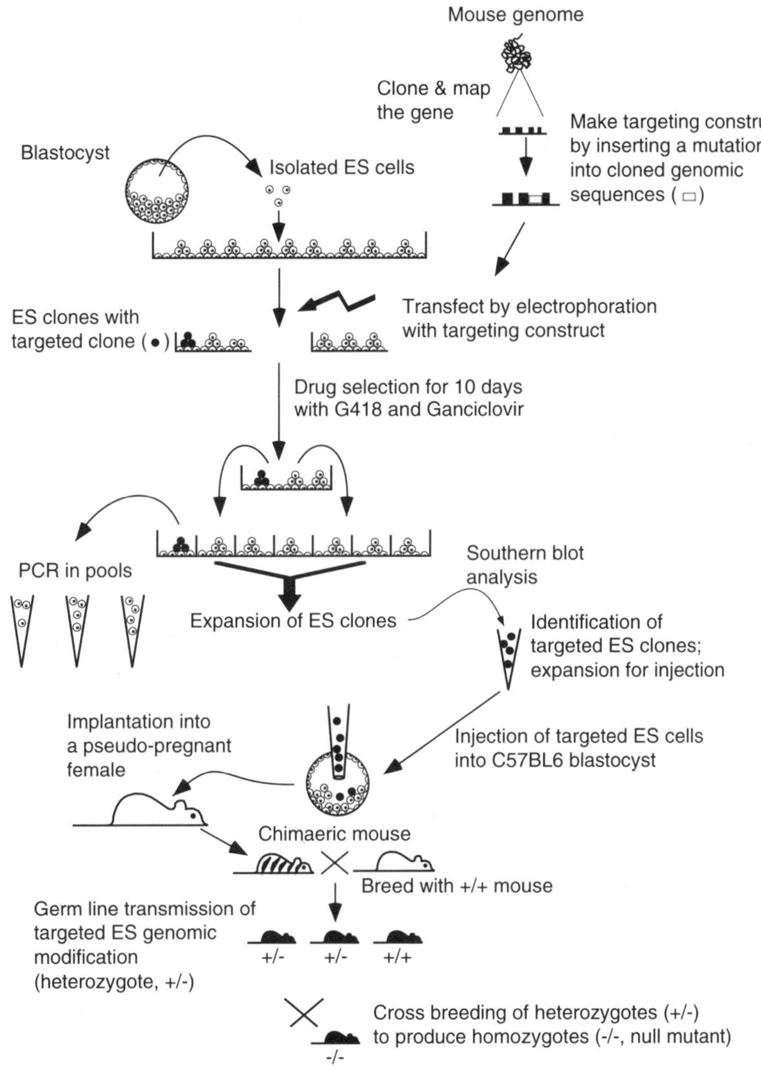

Figure 33.2 Schematic diagram outlining the strategy used to generate gene targeted (knock-out type) in mice. Embryonic stem (ES) cells derived from 129SV mouse blastocysts are transfected with a targeting construct and placed under drug selection for 10 days. ES cell clones that survive this period of selection are then picked and grown up in separate wells. A portion of these clones are then examined for homologous recombination (targeted) events by analysing the DNA of these ES clones by polymerase chain reaction (PCR) assay and southern blot. Once identified, targeted ES clones are expanded for injection into C57BL6 mouse blastocysts. These chimaeric blastocysts (129SV+C57BL6) are then placed into pseudo-pregnant females. Chimaeric embryos are subsequently born and are then bred with wild type mice for germ line transmission of the targeted ES cell genomic modification (heterozygote +/). Heterozygotes are finally cross bred to produce homozygote mice (–/– null mutant).

to test for the desired intergration (see Figure 33.2), and secondly a detailed analysis of the endogenous gene's DNA by southern blot is performed. Southern blot involves restriction digestion of the whole genome, separation of the restriction fragments by electrophoresis and probing of the various fragments with a radioactive homologous sequence to test for the presence of the HR event (see Chapter 26).

Stage 6: Expansion and injection of mutated ES cells into developing embryos

ES cell clones that have been positively identified as having undergone an HR event are then expanded up for injection. At this stage, the ES cells have under gone a lot of stress in culture and may not be in the best shape. A proportion of the ES cell clone may have under gone differentiation, chromosomal translocations and/or deletions. Cell differentiation can be kept at a minium by maintaining strict ES culturing techniques and working with a low ES passage number. Chromosomal abnormalities can be checked by doing chromosomal spreads (Hogan et al., 1994) and having them examined by an expert for gross abnormalities. Most investigators these days don't worry too much about this, as any gross chromosomal rearrangement would likely result in failure of the embryo to develop.

Subsequent breeding will also go some way to segregating non targeted mutations from the targeted mutations (unless the mutation is close to the locus of the endogenous gene, i.e. it is linked). Nevertheless, it is important that gene targeted mice be generated from more than one mutant ES cell clone.

ES cells are injected into developing mouse embryos at the blastocyst stage of development, 2.5 day after fertilisation. Females are super ovulated with hormone injections and mated with healthy male studs. These pregnant females are sacrificed and blastocysts are removed and placed into culture dishes with the appropriate buffer ready for the injection with mutant ES cells. Mutant ES cells are introduced into the blastocysts via an injection pipette (Figure 33.2). While these cells are being loaded into the injection pipette, a blastocyst is picked up with a holding pipette ready for stabbing with the injection pipette. In a typical injection session, an experienced worker will have injected some 60 to 100 blastocysts with manipulated ES cells.

Once the injections of mouse blastocysts are complete, chimaeric embryos are then transferred into pseudo-pregnant females' uterus for further development and eventual birth (Figure 33.2). Typically 8 to 10 chimaeric embryos are placed in the uterine horns of these pseudo-pregnant females (4 to 5 per uterine horn). These females have been made pseudo pregnant by prior mating to vasectomized males.

In order to track the contribution of ES cells in the developing tissues of the host embryo, a different strain of ES cells to that of the host embryo is used. For example, most ES cells used today are derived from a 129SV mouse that carries agouti (yellow-brown) coat color, and the popular choice of host embryo is a C57BL6 mouse that has a black coat color. The chimaeric mouse that is derived from donor ES 129SV and host C57BL6 cells will have a black and agouti coat color. In theory, the mix of cells seen in the coat color is representative of the relative contribution of these cells to the other tissues that make up this chimera. In practice however, the percentage of ES cells that make up the germ cell population is much lower than the coat color would suggest. Thus for ES cells to effectively populate the germ cell population, the chimaera should be about 80% agouti (80% donor derived). Although some investigators believe that even with low percentage agouti coat color some ES cells will contribute to the germ cell population.

Stage 7: Interbreeding of offspring to inactivate the second copy of the gene—homozygote mutant

Once manipulated ES cells populate the germ line population of the chimaeric embryo, it is possible to stably transfer this mutation via the gametes into subsequent generations of mice. The reason for doing this is that gene targeting of ES cells *in vitro* usually only inactivates one copy of the targeted gene. Transferring it via the gametes produces mice with one stable copy of the mutant gene (heterozygote). Interbreeding of heterozygotes will produce mice that have two copies of the mutated gene (homozygotes). Typically, the mating of two heterozygotes produces a litter of half wild type (no copies of the mutant gene), a quarter heterozygote and a quarter homozygote (Figure 33.2). In order to genotype these offspring, analysis of their gDNA is required. Most workers take a small amount of tissue (usually a bit of tail or ear) digest it, and analyse the DNA by either PCR or southern blot for the presence of the mutated gene. Usually the same or a similar strategy for detecting HR events in ES cell clones is used to determine the genotype.

Stage 8: Analysis of mutant mice

Analysis of mutant mice roughly falls into the following sections: (i) ensuring that the gene has been targeted successfully; (ii) checking for pathological abnormalities from gross though to microscopic; (iii) rescue of the phenotype; (iv) further studies to determine the precise role of a particular molecule in the animal's development and function.

Through the genotyping of the offspring, investigators are able to determine if the targeted mutation has occurred. To then check if this mutation has resulted in altered mRNA being transcribed, northern blot analysis of the mRNA is usually done. In addition, sequencing of the mRNA will determine whether it is junk or functional. Apart from sequencing the mRNA, the best and easiest method to determine if a function protein is being translated from the targeted gene is through the use of antibodies directed against the molecule normally encoded by the gene. Ideally it is best to have antibodies that recognise both ends of the molecule (the N and C terminus respectively). Western blot protein analysis or immunohistochemistry of tissues known to express the molecule are undertaken to look for the absence or presence of the molecule or even a truncated form of the molecule (see Chapters 19 and 25).

Once the absence (or altered form) of the targeted molecule has been established, investigators then set about on a pathological analysis to determine the phenotype. Phenotypes can range from lethality to no apparent abnormality. A thorough pathological analysis should include gross anatomical investigation, sample light microscopy of these tissues, followed by a detailed analysis of tissues that are known to express the targeted molecule.

One way to assess if altered phenotype is due to the inactivation of the targeted molecule is to perform a rescue experiment. The easiest way to do this is to inject the cDNA of the particular gene into the pronuclei of a homozygote mutant fertilised egg. In theory, this should produce a transgenic mouse containing a normal cDNA of the targeted gene. Another, but less direct, way is to isolate cells from the homozygote mutant that would normally express the molecule of interest and reintroduce by transfection a functional copy of the gene. For example rapsyn, a 43kDa cytoskeletal molecule is known to cluster nicotinic acetylcholine receptors (AChRs) *in vitro*. In rapsyn deficient mutant mice, AChRs fail to cluster on the surface of muscle cells. Taking these rapsyn deficient muscle cells, we grew them in culture, and transfected them with the rapsyn cDNA. Rapsyn was expressed, and as a result AChR clustering was restored (i.e. we had reversed the phenotype; Han et al., 1997).

The direction for further analysis is largely determined by what is already known about the molecule and the other molecules with which it interacts with. Such studies may focus on working out the molecular pathway that a particular molecule is part of (cell biology) or focus more on the functional aspects in the animal or tissues (physiology).

All analysis should be done on litter matched pairs and should include wild type, heterozygote and homozygote mice. Comparison between all three genotypes is necessary and will provide information to the investigators as to whether there are differences between homozygote compared to wild type and heterozygote (a recessive phenotype), or that both the homozygote and heterozygote differ form the wild type (a dominant phenotype).

POTENTIAL PROBLEMS AND FUTURE DIRECTIONS

Mixed genetic background

One disadvantage of generating a chimaeric mouse composed of two strains is a mixed genetic background. Subsequent interbreeding of the chimaera's offspring will retain this mixed genetic background unless at some point the offspring are bred with the same strain of mouse that the ES cells came from, in this case 129SV mice.

Why is this important? To be confident that the phenotype observed in a mutant mouse is a result of mutation to the targeted gene and not some other mutation in the genome, a uniform genetic background is a major advantage. Successive breeding in a constant genetic background with the same phenotype occurring in subsequent generations gives the researcher the confidence that the abnormality is due to the mutation alone. Even for a mixed genetic background, successive breeding should segregate out other mutations that could exist in the genome provided that the mutation is not linked (i.e. on another chromosome or greater than about 16 centiMorgans away from the targeted gene). Sixteen centiMorgans of gDNA represents about 300 genes (Gerlai, 1996 and references therein). This could be a problem if the gene being targeted is a member of a gene family that is not randomly distributed throughout the genome but is contiguous on one chromosome, as is sometimes the case for families of related genes.

One obvious way around this problem is to breed the progeny of the chimera with 129SV mice. Unfortunately, the 129SV mouse is not an ideal inbred strain. It breeds very poorly and has a lot of spontaneous mutations in its genome which seem to affect the function of its nervous system. This latter issue can be a worry to neuroscientists, especially when studying complicated phenotypes such as memory and learning in mutant mice that may be due to the function of a number of co-ordinately expressed genes.

These issues are now being discussed between molecular biologists, geneticists, mouse breeders and neuroscientists (TINS 1996, Vol. 19(5)), and there are a number of solutions to this problem. First, as mentioned previously, the progeny of the 129-C57BL6 chimera can be bred (back crossed) with C57BL6 mice, and in this case, eventually all of the genome will be C57BL6, but the targeted gene and surrounding gDNA (about 16 centiMorgans; Festing 1992; Gerlai, 1996) will always be 129SV. Another solution is to generate other strains of ES cells in combination with other strains of host embryos. This is currently under way in a number of laboratories, but has been technically difficult. Nevertheless these issues will be over come and the next generation of ES derived transgenics will be on suitable genetic backgrounds.

Lethal or no phenotype

Another potential problem is that inactivation of a targeted gene may produce a mutant mouse with no apparent phenotype, or be lethal at a very early stage of development before the stage of interest. Once again, there are a number of potential ways around these problems, and these are currently being explored. For example, one way is to be able to have the molecule expressed normally but then to be able to switch it off in particular tissues or at particular stages of development.

To achieve this, investigators are now generating tissue and inducible gene targeted mice, using the Cre-*lox*P system. In nature, this system operates when *Escherichia coli* is infected by a virus bacteriophage P1. During infection the enzyme Cre-recombinase from bacteriophage P1's 'job is to separate any phage genomes that become joined to one another. To do that, Cre lines up short sequences of phage DNA called *lox*P sites and removes the DNA between them, leaving on *lox*P site behind' (Barinaga, 1994). Brian Sauer first showed that this system could be made to work in eurkaryotic cells (see Baringa, 1994), and recently Rajewsky and colleagues have extended this system to gene targeting in mice (Gu et al., 1994).

Briefly, this involves first making a pronuclei transgenic, where a tissue specific or inducible promoter is linked to a bacterial enzyme Cre-recombinase. In such a mouse, the Cre-recombinase enzyme is only expressed in specific cells. These mice develop normally as the target for this enzyme *lox*P is not present in the animal. In the second animal, a standard gene targeted mouse is made carrying the targeted gene that is flanked by *lox*P elements and the neo gene. This is achieved by designing a targeting construct that has the *lox*P elements outside the gene or in intronic regions that flank coding exons, and the neo gene cassette is inserted outside the gene (or in intronic sequences). This is important, as expression of the neo and the presence of the *lox*P elements must not interfere with normal expression of targeted gene. The transgenic Cre mouse is then crossed with a homozygote *lox*P neo gene targeted mouse. Both are then cross bred. In cells where Cre is expressed it will bind to its target *lox*P and excise gDNA sequences between the *lox*P sites, thus effectively removing the gene in cells where Cre is being expressed (or induced).

The obvious question many neuroscientists are now asking is whether this technology can be transferred to other animals such as rats, pigs, and so forth. In theory, it is possible and there have been reports of ES cells derived from rats and guineapigs, but no great advances have been made despite intense efforts by many investigators. Perhaps the best strategy is to improve the frequency of homologous recombination (the basis of gene targeting) to a point where it can be done directly by injecting the targeting construct into either the oocyte (female egg) or into the pronuclei of a fertilised egg.

Whatever the future modifications, gene targeted transgenics are likely to become a powerful tool gaining insights into the genetic basis of how we think and interact with our environment.

REFERENCES AND FURTHER READING

Ausubel, F. M., Brent, R. Kingston, R. E., Moore, D. D., Seidman, J. G. Smith, J. A. and Struhl, K. (1992) *Short Protocols in Molecular Biology* (2nd edn). New York: Greens Publishing Associates and John Wiley and Sons.

Barinaga, M. (1994) Knockout mice: round two. *Science* **265**, 26–28.

Bradley, A., Evans, M. J., Kaufman, M. H. and Robertson, E. J. (1984) Formation of germ-line chimaeras from embryo-derived teratocarcinoma cell lines. *Nature* **309**, 255–256.

Capecchi, M. R. (1989) The new mouse genetics: altering the genome by gene targeting. *Trends in Genetics* **5**, 70–76.

Evans, M. J. and Kaufman, M. H. (1981) Establishment in culture of pluripotential cells from mouse embryos. *Nature* **292**, 154–156.

Festing, M. F. W. (1992) From character to gene some strategies for identifying single gene controlling behavioural characteristics. In *Techniques for the Genetic Analysis of Brain and Behavior: Focus on the Mouse*, edited by D. Goldowiz, D. Wahlsten and R. E. Wimer, pp 17–38. Amsterdam: Elsevier.

Gerlai, R. (1996) Gene-targeting studies of mammalian behavior: is it the mutation or the background genotype? *Trends in Neurosciences* **19**, 177–181.

Gu, H., Marth, J. D., Orban, P. C., Mossmann, H. and Rajewsky, K. (1994) Deletion of a DNA polymerase gene segment in T cells using type-specific gene targeting. *Science* **265**, 103–106.

Han, H., Noakes, P. G. and Phillips, W. D. (1997) Recombinant rapsyn restores acetylcholine receptor clustering to mutant myotubes. *Proc. Aust. Neurosci.* **8**, 6–12.

Hasty, P. and Bradley, A. (1992) Gene targeting vectors for mammalian cells. In *Gene Targeting* (The Practical Approach Series), edited by A. L. Joyner, pp. 1–30. IRL Press.

Hogan, B., Beddington, R., Costantini, F. and Lacy, E. (1994) *Manipulating the Mouse Embryo: A Laboratory Manual* (2nd edn). New York: Cold Spring Harbor Laboratory Press.

Mansour, S. L., Thomas, K. R. and Capecchi, M. R. (1988) Disruption of the proto-oncogene *int-2* in mouse embryo stem cells: a general strategy for targeting mutahons to non-selectable genes. *Nature* **336**, 348–352.

Martin, G. R. (1981) Isolation of a pluripotent cell line from early mouse embryos cultured in medium conditioned by teratocarcinoma stem cells. *Proc. Natl. Acad. Sci.* **78**, 7634–7638.

Robertson, E. J., Bradley, A., Kuehn, M. and Evans, M. (1986) Germ-line transmission of gene introduced into cultured pluripotential cells by retroviral vector. *Nature* **323**, 445–448.

Sambrook, J., Fritsch, E. F. and Maniatis, T. (1989) *Molecular Cloning: A Laboratory Manual* (2nd edn). New York: Cold Spring Harbor Laboratory Press.

SECTION 13

NEURAL GRAFTING

CHAPTER 34

NEURAL GRAFTING

L. J. Reece
Developmental Neurobiology, Research School of Biological Sciences, Australian National University

INTRODUCTION

The concept of transferring parts of the body from a donor to a host, or *transplantation*, has become an everyday occurrence around the world for organs such as the kidney and cornea. Hopes have been raised in recent years for application of similar techniques to transplantation of parts of the brain from one individual to another for repair of damaged circuitries. Between the years of 1970 and 1975, several investigators grafted fetal rat neurons into older rats and used a combination of histochemical, autoradiographic, and electrophysiological methods to evaluate the properties exhibited by transplanted neurons. These studies established firmly that such mammalian grafted neurons will survive for long periods of time in host tissue. Although the technique is not in regular use clinically, neuronal transplantation has since become a popular method for exploring a variety of experimental questions in the mammalian nervous system as well as carrying with it great hopes for recovery of nervous system function in humans.

Although the earliest studies utilized blocks of tissue implanted into cavities or troughs in the cortex, many other methodologies have since emerged. Among them are dissociated cell suspensions, fragmentation, intraventricular placement, and intraocular grafts. Each of these has advantages and disadvantages; however, for many applications, stereotaxic injections of cell suspensions has become the method of choice and will be described here. This application combines tissue culture techniques and the use of *en bloc* tissue transplantation; the method is constantly being refined as the broad area of tissue culture techniques and genetic manipulation is advanced.

DEFINITIONS

Neural grafting, as neural transplantation is also known, involves the transfer of some type of neural tissue from one place to another. The source of the tissue is termed the *donor*; the recipient is known as the *host*. The donor and host may be the same individual (*autograft*), they may be the same species but different individuals (*allograft*), or they may be from different species altogether (*xenograft*). Recent advances in immunology and genetics mean that separate individuals from the same litter may have the same genetic make-up, rendering such tissue to be virtual autografts.

Allografts have until recently been the most frequently used type of transplant for experimental purposes. Fetal material harvested from litters of rats, cats, and individual monkey embryos has been widely used for elucidating the mechanisms by which neural grafts can ameliorate nervous system dysfunction. Xenografts have probably the least chance of survival in most cases, but they do offer some experimental advantages such as providing an immunological means of labelling graft cells. In addition, grafts between species may offer the most hope for future clinical use, since genetic engineering techniques permit manipulation of nuclear material in donor cells thus affecting the resultant phenotype expression and material produced.

The advantage of autografts lies in their similarity, genetically and immunologically, to the host tissue. Although the brain has been traditionally considered to be an 'immunologically privileged' site, we now know that neural grafts can and do undergo rejection similar to that with larger organ transplants. That probability of rejection is probably increased by increased genetic differences. Drawbacks are mostly of clinical relevance—there is little purpose clinically in destroying one structure to replace another, and if the structure of interest is damaged, tissue must come from another individual.

TECHNIQUES

Briefly, the technique involves choosing an appropriate (for the question being answered) host and donor, preparation of the cell suspension on the day of grafting, and stereotaxic injection of the cells. There are as many variations of this technique as there are CNS tissue types, and investigators are constantly trying to improve the specificity of cell type via new cell sorting and genetic engineering methods.

Selection of donor and host

Major considerations contributing to graft survival appear to be the following:

1. Donor age
2. Mechanical handling
3. Presence of necessary trophic factors
4. Host age

Donor age is apparently the most important. Age plays a role where the cells have extended neurites. Preparation of the suspension axotomizes the cells; those which tolerate suspension best are at an age just prior to final differentiation or which have just differentiated but not extended neurites. For suspensions, the optimum neuronal survival is coincident with periods of cell differentiation, proliferation, and migration; these parameters vary with each brain region, and within each region, each cell type. (For a good summary, see Table 1 in Chapter 1 of Dunnett and Björklund (1992).) For animal experimentation purposes, these tissues are normally obtained from pregnant animals where the time of mating is known to within 12–24 hrs (*time-mated*).

Concerning choice of host, grafts into immature hosts, where the influences strongest in development (such as trophic factors like nerve growth factor) are still present, are also the most successful. Trophic factors or their receptors often appear transiently during development, then disappear when the growth process which they affect is completed. Disappearance of such factors appears to be the reason that grafts into or from older animals are generally not as successful. Much of the recent experimental work on duplication of neural circuits has emphasized the necessity for re-creating the biochemical environment present during original development of the tissue. Where the host environment is supplemented with growth-enhancing factors (trophic), the rate of survival is generally increased (Takayama et al., 1995).

Grafts into adult hosts, however, have a reasonable rate of success and yield information about mature target influences on ingrowing axons. Animals of species ranging from frogs and salamanders to macaques and man have undergone neural transplantation; the procedures described here are for the rat, but (with some restrictions) are generally applicable to all mammalian species.

Adult hosts should generally undergo one of several types of lesions prior to grafting; these include injection of neuro- or excito-toxins, radio-frequency lesions, or aspiration. Lesions are thought to participate in graft survival in several ways, including stimulation of trophic factors and creation of 'synaptic space' to harbor ingrowth of new connections. Where such lesions are not present, the grafted tissue is often out-competed by intrinsic neural fibers for connections and remain isolated and poorly integrated into the host. In addition, lesions provide a deficit—behavioral, histochemical, electrophysiological or otherwise—the alteration or improvement of which can be measured to assess the successful function of the graft. Much of the controversy involved in reporting graft success relates to the assessment methods employed; increases in chemical levels of neurotransmitters or increased cell counts may not translate to actual physiological function of grafted tissue.

Preparation of cell suspension

The actual preparation of tissue from donors and injection into the host requires speed and efficiency; adjacent work stations for host and donor so that tasks can be dovetailed is convenient (Figure 34.1). Measure and record the crown-rump length (CRL) of the embryo for later reporting; CRL is a good estimate of embryonic age. Where age is absolutely critical, time-mating the animals to a specific window of time (12–24 hr) is better.

A variety of sterile media based on tissue culture agents or common balanced salt solutions are now used for cell suspensions. Selection depends very much on the requirements of the cells of interest. There are generally two procedures for obtaining pieces of fetal tissue from an anaesthetized, time-mated pregnant rat. The first is to remove fetuses beginning at the distal end of the uterine horn by first clamping with hemostats just proximal to the most distal fetus to prevent bleeding. The amniotic sac is then punctured to permit a small amount of sterile amniotic fluid to drip onto a sterile slide or small dish. Trying not to touch the tissue with non-sterile instruments or hands, the embryo is squeezed out of the sac onto the fluid area of the slide and removed for dissection. The second method is to remove the whole uterine horn after clamping close to its source (one arm of the Y-shaped uterus), and remove the whole horn to a sterile dish full of sterile media or saline-glucose solution. Remove each embryo as above, then proceed to the second horn. Once all the embryos have been removed, the mother should be overdosed with anaesthetic.

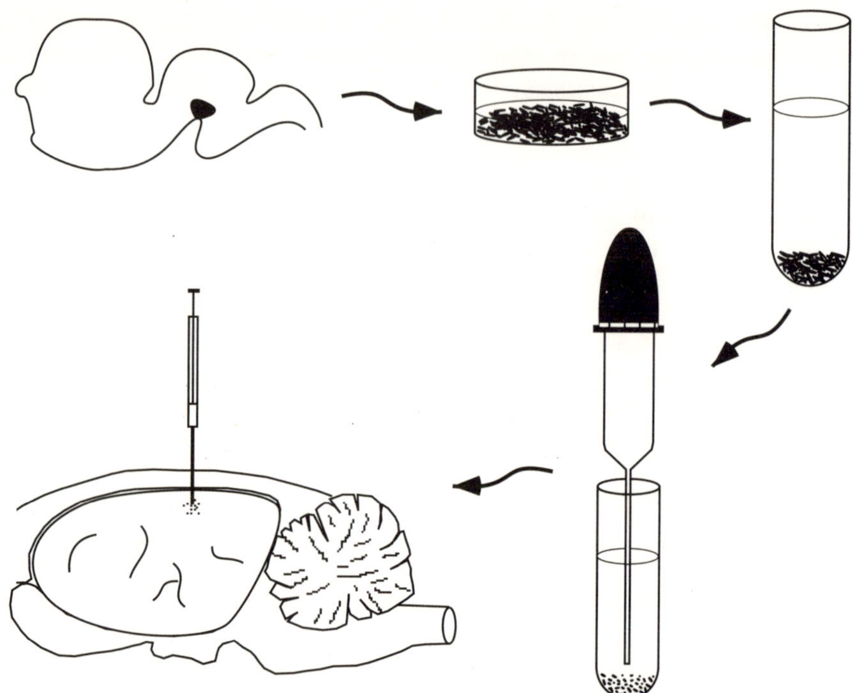

Figure 34.1 Five step procedure for cell suspension transplantation. Clockwise from upper left: dissection of embryonic area of interest; inspection of brain pieces; incubation and washing collected tissue; mechanical break-up of tissue pieces to form cell suspension; injection of cell suspension into host target area.

Dissection

Dissection of different brain regions varies with anatomical region and laboratory regimes. There are too many variations to go into here, but good references are found in Dunnett and Björklund (1992). Dissection is aided by transillumination of the embryo by a fiber optic light. If normal lighting is used, a black background aids in discriminating crucial brain features. Collect dissected fetal brain tissues in a sterile dish of main media over 20–60 minute period. Speed in handling improves the graft survival rate. The ingredients in the media will depend much on the brain region desired. There are several brain regions whose cells do not tolerate a protein digestion step which is often included to aid in separating the cells, including some cells in the locus coeruleus and substantia nigra.

Add desired volume (for injection) of solution for final tissue suspension (usually 5–10 μl for each fetal brain piece). Carefully triturate (agitate) with a series of fire-polished pasteur pipettes with openings of decreasing diameter for 2–4 minutes until well-dissociated. Avoid making bubbles. The suspension should be slightly cloudy and have no large clumps of tissue when held up to a bright light or optical fiber light.

It is often desirable to check cell viability and the extent of cell dissociation. Several histochemical methods are available for this, including Trypan blue exclusion, acridine orange or ethidium bromide. Healthy cells will not take up acridine orange, but they do take up ethidium bromide and will fluoresce green when stimulated by long wavelength UV light.

Stereotaxic injection

Since the successful grafting of neural tissue often depends on the accuracy of the implantation in the host, such transfer should be done with as much precision as possible. This usually involves utilising direct surgical visualization, or where necessary, stereotaxic injection. The host should be anaesthetized and mounted in a stereotaxic apparatus (see Chapter 15). Locate the bregma (the joining of the sagittal suture with the coronal or transverse suture); most stereotaxic atlases give co-ordinates with reference to bregma. Mark the location with a pencil or felt tip marker. Use a dental drill to make pilot holes for injection.

After doing a last trituration, draw up the desired amount of suspension in a tube attached to a Hamilton syringe. In order to inject most of the suspension, it is necessary to use the syringe as a displacement measuring device for the suspension, but not to contain it, since too much of the suspension is left in the syringe. Use of too small a needle will also damage the cells.

Take the needle tip down to the level of the dura mater, and make a stereotaxic measure for zero depth. Depth of injection is critical when injections are made blind (without seeing the injection site). Then descend slowly to a depth in accordance with placement of the appropriate structure. Inject the suspension at a rate of approximately 1 μl/min. Too rapid a rate will cause suspension to travel back up the needle path; a slow injection permits the surrounding host tissue to adapt to the pressure. Once the desired volume is injected, wait 5 minutes to allow the suspension to permeate surrounding tissues; then raise the needle slowly. Rapid removal will suck part of the cells up along with the needle. Pack the hole with moist Gelfoam and suture the skin over the top, treating the skin edges with antibacterial agent.

Allow the animal to recover. Time and method of graft assessment will depend upon the parameter of interest. Sham operations and controls with normal saline or main media injections should of course be included in any experimental paradigm.

OTHER CONSIDERATIONS

Care must be taken to minimize both mechanical trauma due to trituration and also bacteriological infection. Although many people claim only minimal sterile techniques are necessary with rats, the incubation step provides idea environment for bacterial growth.

Since the injected suspension has cells which are still proliferating, the mature volume of the graft may be many times the volume injected. Care must be taken to adjust the injection volume to the proliferative capacity of the brain area of interest, which varies markedly according to the age of the embryonic donor. If the total volume is too great increased intracranial pressure may cause extrusion of the graft back up to the cortical or ventricular surface, far away from its desired location.

Immunosuppression, common in other organ transplantation to prevent tissue rejection, has been demonstrated to increase survival of neural grafts in many cases as well (DiLoreto et al., 1996). Cyclosporin A given daily is commonly used.

EVALUATION

Graft success is assessed in a variety of ways; of course in a clinical situation the most important property is return of function, especially where cognitive abilities have been impaired. Judging such return in small animal experimentation is often difficult and function must be extrapolated from other more indirect tests. Many techniques are used for estimation of return of function; the basis of many are described elsewhere in this guide.

CONCLUSION

The research which has followed pioneering studies in neural grafting has been aimed in large part at studying whether connections between the host and graft are functional and appropriate, and at determining what factors govern establishment of those connections in each CNS system. The possible clinical applications of such techniques has not gone unnoticed; the area of neuronal grafting is being explored by many with the hope of reconstructing damaged circuitries within the brain and spinal cord. However, neural grafting is a powerful tool in itself for investigating questions of development and regeneration. Combined with new techniques which permit increasing specificity of cell types, altering surface molecules, and neurotransmitter content, the methods offer an excellent opportunity for the study of neuronal interactions during growth and development.

REFERENCES AND FURTHER READING

Cheng, H., Almstrom, S. & Olson, L. (1995) Fibrin glue used as an adhesive agent in CNS tissues. *J. Neural Transplant. Plast.* **5**, 233–43.

Cheng, H., Cao,Y. & Olson, L. (1996) Spinal cord repair in adult paraplegic rats: partial restoration of hind limb function. *Science* **273**, 510–3.

DiLoreto, D. Jr., del Cerro, C. & del Cerro, M. (1996) Cyclosporine treatment promotes survival of human fetal neural retina transplanted to the subretinal space of the light-damaged Fischer 344 rat. *Exp. Neurol.* **140**, 37–42.

Dunnett, S. B. and Björklund, A. (1992) *Neural Transplantation: A Practical Approach*. Oxford: IRL Press.

Peschanski, M., Cesaro, P. & Hantraye, P. (1995) Rationale for intrastriatal grafting of striatal neuroblasts in patients with Huntington's disease. *Neuroscience* **68**, 273–285.

Takayama, H., Ray, J., Raymon, H. K., Baird, A., Hogg, J., Fisher, L. J. and Gage, F. H. (1995) Basic fibroblast growth factor increases dopaminergic graft survival and function in a rat model of Parkinson's disease. *Nat. Med.* **1**, 53–58.

GLOSSARY

Acoustic startle Innate reflex in which an animal 'jumps' in response to a loud noise.

Active transport Transport of a solute or particle requiring energy expenditure by the cell.

Acutely dissociated neurons Neurons that have been obtained by dissociating nervous tissue into single cells, usually by a combination of enzyme treatment (e.g. with trypsin) and mechanical agitation. The dissociation is commonly done just before the neurons are needed; that is, they are not maintained in culture medium. (See cultured dissociated neurons.)

Adjuvant A compound (often Freund's adjuvant) that enhances the immune response.

Afferent Conveying towards; nerve pathways coming before.

Affinity chromatography A technique in which molecules are separated by their relative ability to interact with a specific immobilized substrate or antibody.

Affinity tag An artificially added sequence to a recombinant protein which encodes a known epitope (or amino acid sequence) so that the resultant protein may be purified or monitored by antibodies that specifically recognize that epitope or tag.

Aliasing The process by which under sampling of analog signals (i.e. at less than the Nyquist frequency) produces spurious, low frequency signals which are not present in the measured signal.

Allograft Tissue or organs taken from one individual and transferred to another individual of the same species; synonymous with homograft.

Amplifier Apparatus used to amplify the electrical signal derived from a microelectrode electrode recording of the transmembrane potential or current.

Analog signal A continuous representation of a signal. This is in contrast to digital signals, where the magnitude is sampled at discrete intervals and is therefore represented as a series of values.

Anchor primer An oligonucleotide that anneals to the 3′ end of the DNA template.

Annealing When double-stranded nucleic acids are denatured (dissociated) by heat and allowed to reform specific duplexes at a temperature just below their melting temperature.

Anterograde Travel in a forward direction, i.e. away from the neuronal cell body.

Antibody A molecule produced by lymphocytes that binds to specific foreign molecules in the body.

Antimitotics Agents that are added to neuronal cultures for a day or two in order to suppress division (mitosis) of glial cells. Without antimitotics, glia will continue to divide and may overwhelm the neurons in the culture. Examples of antimitotics are Ara-C (cytosine arabinoside) and FdU (fluorodeoxyuridine).

Antisense string A string of nucleotides that does not code genetic information.

Antiserum Serum prepared from immunized animals and containing unpurified polyclonal antibodies.

Arbitrary primer An oligonucleotide that anneals to the 5′ end of the DNA template.

Artificial cerebrospinal fluid (ACSF) A solution whose ionic composition is designed to mimic the ionic composition of cerebrospinal fluid (CSF).

Artificial membrane A mixture of phospholipids dispersed in a solvent and thinned to form a bilayer.

Autofluorescence Intrinsic fluorescence emitted by cell structures when they are exposed to certain wavelengths of light.

Autograft Tissue or organs taken from one part and transferred to another part of the same individual.

Autoradiography Technique used to visualize and/or quantify the distribution pattern of radiolabelled cellular constituents.

Axotomy Severing of the axon of a nerve cell; can result in degeneration or cell death.

Bacteriophage A virus that lives within a bacterium such as *Escherichia coli* (*E. coli*).

Bandwidth (of light) A continuous range of wavelengths (or frequencies) of light is termed a bandwidth e.g. fura-2 has an excitation bandwidth of 332–420 nm.

Bessel filter A type of low pass filter characterized by a broad cut-off region in its frequency response above its pass-band.

Binary code The code by which computers deal with information. Each piece of information (Byte) is stored as a series of zeros and ones (Bits).

Bipolar electrode One of two recording electrodes usually placed equidistantly from the biological signal of interest. Surface EMG electrodes are examples of bipolar electrodes.

Blastocyst A mouse embryo at 2.5 days after fertilisation at the stage of implantation into the uterine wall. It is spherical in shape and its surface is called the trophectoderm. About half of the interior space of the sphere is filled with cells (the rest is hollow) and these are referred to as the inner cell mass.

Bregma Joining of the sagittal suture with the coronal or transverse suture in the skull.

Bridge (mode or balance) The electronic circuitry used to compensate for the voltage drop associated with passage of current through a microelectrode, and thus ensure correct measurement of membrane potential.

Butterworth filter A type of low pass filter characterized by a sharp cut-off in its frequency response above its pass-band.

Cannula A small tube for insertion into the brain (or blood vessel) through which agents can be infused.

CCD camera Camera based on an array of silicon photodiodes or charge coupled devices.

cDNA Complementary DNA; double-stranded DNA which is made from a messenger RNA (mRNA) template and therefore is complementary to that mRNA.

Cell-attached patch An 'on-cell' patch-clamp recording configuration, in which the small patch of cell membrane isolated by the patch electrode remains part of the cell.

Cell line A class of cells that will continue to divide in culture indefinitely. They are usually derived from tumours. Examples of neuronal cell lines are neuroblastoma cells and pheochromocytoma (PC12) cells.

CentiMorgan A unit of distance along chromosomal DNA that gives a recombination frequency of 1%.

Chimaera Single organism developing from an embryo that is composed of cells from two different individuals and thus of two different genotypes.

Chromosomal spreads Cells are lysed at a stage when they are undergoing cell division, the chromosomes are released from the nucleus, fixed, stained and examined.

Chromosomes Rod shaped bodies visible in the eukaryotic cell nucleus during cell division. They contain the genes in linear order. Each chromosome consists of a very long molecule of DNA associated with proteins.

Chronic Marked by long duration; permanent.

Complementary strands Two nucleic acid strands containing linear sequences of nucleotides whose bases specifically pair by Watson–Crick base pairing. Strands are anti-parallel: one chain runs $5' \rightarrow 3'$ and the other runs $3' \rightarrow 5'$.

Conditioned response A response elicited by a stimulus through association with another stimulus.

Confocal laser scanning microscope Specialized microscope, using a laser as a light source, that is capable of producing high resolution images.

Corner frequency The frequency at which the filter frequency response falls by 50% (–3dB).

Cross-correlogram This shows the temporal relationship between the occurrences of spikes in one neuron (or motor unit) and those in the another neuron (or motor unit). Cross-correlograms can give insights into whether one neuron (or motor unit) is excited or inhibited by the other or both respond to a common input from a third source.

Culture medium Sterile solution used for maintaining cultured neurons. For mammalian neurons, it typically comprises a bicarbonate-buffered high-sodium salt solution, essential amino acids and vitamins, glucose, antibiotics and fetal bovine serum (FBS). Serum-free variants of the culture medium substitute various hormones, growth factors, etc. for the FBS. To maintain the correct pH and growth conditions, cultures are usually kept in a 37°C incubator that is gassed with 5% CO_2.

Cultured dissociated neurons Neurons that have been dissociated from nervous tissue into single cells under sterile conditions, then maintained in culture medium in an incubator for a day or more (up to several weeks). The dissociation is normally done using a combination of enzyme treatment (e.g. with trypsin) and mechanical agitation. Dissociated cultures can form functional synapses.

Current-clamp A name given to the electronic circuitry typically used to measure transmembrane potentials. The circuitry is designed to ensure that the current through the microelectrode is held steady (or at zero if desired).

Dark adaptation As the eye passes from light conditions to dark so there is an increase in the electroretinogram (ERG) sensitivity. This reflects the time taken to regenerate rhodopsin levels in the photoreceptor outer segments and a resetting of the gain mechanism of the phototransduction biochemical cascade. If the ERG to a brief light stimulus of maximum intensity is recorded intermittently over a period of time, the response amplitude gradually increases and eventually saturates (e.g. after 90 minutes). Plotting the resulting amplitude versus time data provides an example of a dark adaptation curve.

Densitometry Quantitative analysis of the intensity of bands on a gel or Southern or Northern blot.

Deoxyribonucleic acid (DNA) Bases that form genes in the nucleus.

Desterification The process of cleaving acetoxymethyl ester groups from a dye molecule.

Dialysis Separation of solutes by diffusion through a semipermeable membrane.

Differentiation The process by which cells divide into cells with a greater level of specialization.

Discontinuous current-clamp The name given to a method of recording transmembrane potentials whereby the single microelectrode rapidly switches between passing current and recording voltage. After passage of current the voltage drop across the microelectrode tip decays according to the time constant of the microelectrode (a product of its resistance and capacitance). As the voltage decay tends towards a plateau, a true measurement of membrane potential is obtained, unconfounded by the voltage drop across the resistance of the microelectrode.

DNA polymerase An enzyme that catalyses the addition of a deoxyribonucleotide to the 3′ end of a DNA chain.

Donor The source of tissue or organs used in transplantation or grafting.

Dura mater Literally 'tough mother', the term refers to the largest of several layers of the dura which encapsulate the brain inside the skull.

Electroblotting Transfer of nucleic acids or proteins from a gel to a nylon membrane which is mediated by electrophoresis.

Electrode coupling The electrical interaction between the two channels of a double-barrelled electrode, defined by $R_c = E_2/I_1$ (where R_c is the coupling resistance when a voltage is recorded in one barrel (E_2) when a current (I_1) is passed down the other barrel). If the tips are completely coupled then the voltage recorded in the two barrels will be identical; the lower the coupling resistance the greater the independence of the voltages recorded in each barrel.

Electroencephalogram (EEG) The electrical activity recorded from electrodes placed on the scalp.

Electromyograph (EMG) The electrical signals generated by contracting muscle fibers.

Electroporation A method of introducing DNA into mammalian cells. Cells are placed into a buffered solution (pH 7.4) containing DNA and a current is passed through the solution. This process creates holes in the plasma and nuclear membrane of the cell, allowing DNA to pass in from the extracellular fluid through to the nucleus.

Electrophysiology Study of the electrical properties of living cells.

Electroretinogram A recording of the massed electrical potential generated by the retina in response to a flash of light.

Emission spectrum The range of wavelengths, or colors, which a molecule emits when it is illuminated with a specific excitation wavelength is called its emission spectrum.

Epitope A chemical structure recognized by a single species of antibody. A given macromolecule may contain many epitopes.

Excitation spectrum The range of wavelengths which will excite a molecule and cause it to fluoresce is called its excitation spectrum.

Excitotoxin A substance which exerts a toxic effect on nerve cells by permitting them to become hyperexcitable.
Exon The coding sequence of genomic DNA. It is a portion of DNA that encodes part of a polypeptide chain.
Expression construct A double-stranded DNA sequence containing the protein coding sequence of a gene attached to promoter/enhancer and other DNA sequences to ensure that the gene is expressed when introduced into a host cell. The expression construct is usually assembled as part of a bacterial plasmid to facilitate the growth of large amounts of the construct.
Expression plasmid A bacterial plasmid containing an expression construct.
Extinction The process of eliminating or reducing a conditioned response through lack of reinforcement.
Fascicle A bundle of nerve fibers (axons) enclosed in a connective-tissue sheath. Large nerves contain multiple nerve fascicles.
Fear conditioning Pairing of a cue (such as a light) with a noxious stimulus (such as footshock) in order to elicit an emotional state of fear in the animal when presented the cue.
Fetal bovine serum (FBS) An essential component of most mammalian culture media, providing many of the trophic factors that seem to be required for the survival of cultured neurons.
Fetus Embryonic animal form.
Filter hybridisation A technique for the immobilization and detection of nucleic acids on a solid support.
Filter wheel A spinning disk with filters inserted into its surface. The disk can be rapidly rotated to insert the desired filter into the optical path.
Fluorescein isothiocyanate (FITC) A commonly used chemical which is conjugated to secondary antibodies and fluoresces in the green wavelengths.
Fluo-3 A non-ratiometric dye for calcium imaging.
Freund's complete adjuvant A water-in-oil emulsion, containing mycobaterial membranes, that induces cell-mediated immunity by attracting macrophages and other immune cells to the site of injection. This boosts subsequent responses to injections of antigen, such as a peptide, in Freund's incomplete adjuvant. Thus, injection of a peptide in Freund's complete adjuvant will generate a large immune response to both mycobacteria and peptide.
Freund's incomplete adjuvant Similar to Freund's complete adjuvant (above) except it does not contain mycobacterial membranes. When an antigen (e.g. a peptide) is administered as a boost injection in Freund's incomplete adjuvant, after an injection in Freund's complete adjuvant some weeks before, the immune response to the peptide is enhanced and not that to the mycobacteria of Freund's complete adjuvant, as in the initial injection.
Fura-2 A ratiometric dye for calcium imaging.
Gamma amino butyric acid GABA) An inhibitory neurotransmitter.
Gametes Sperm and ova.
Ganzfeld A ganzfeld photostimulus is one in which light is scattered in all directions (e.g. within a ganzfeld diffusing sphere) such that light enters the pupil of the eye

from all directions and achieves homogeneous stimulation of the entire retina ('full-field' stimulation).

Gel electrophoresis A range of methods, based upon application of a voltage across a gel matrix, used to separate proteins (commonly in accordance with molecular weight) or nucleic acids.

Gel filtration A chromatographic technique, employing a porous stationary phase, which separates molecules on the basis of their relative molecular masses.

Genome The genetic make-up of a single cell or a living organism.

Genomic library DNA library consisting of fragments of chromosomal DNA.

Genotype The genetic make-up of an organism.

Germ cells Population of cells from which gametes (sperm and ova) are generated.

Graft Transferring tissue or organs from donor to host; used in non-clinical settings; interchangeable with transplantation.

Hemostat A surgical clamp often used to control bleeding.

Heterograft Tissue or organs taken from one individual and transferred to another individual of a different species; synonymous with xenograft.

Heterologous system A biological system whose components originate from different animal types. For example, expression of a human protein in *Escherichia coli* (*E. coli)* or in oocytes or in a non-human cell line constitutes an heterologous system.

Heterozygote An animal that has inherited two different versions of a particular gene. Mice are diploid organisms, that is they contain two set of chromosomes, which means the organism contains two copies of a particular gene one copy inherited from each parent.

High-Performance Liquid Chromatography (HPLC) Automated chromatography in which the mobile phase is actively pumped through a column (stationary phase) under relatively high pressure using specialized pneumatic or piston pumps.

Histological techniques Techniques used to fix and stain tissues in order to observe their minute structure with a microscope.

Homograft Tissue or organs taken from one individual and transferred to another individual of the same species; synonymous with allograft.

Homologous recombination Recombination event between a cloned fragment of genomic DNA with the same DNA sequences of the endogenous gene.

Homozygote An animal that has inherited two identical copies of a particular gene.

Host The recipient of the tissue or organs being transplanted or grafted.

Hybridization Formation of double strands of nucleic acid from single strands.

Hybridoma Immortalized cell line derived from the fusion of an antibody-producing B-cell and a myeloma cell.

Hydrogen bond A relatively weak type of bond formed between electronegative atoms in one molecule and hydrogen atoms of another molecule.

Hydrophobic Interaction Chromatography (HIC) A technique which is used to separate molecules on the basis of their relative hydrophobicity. Generally, a decreasing salt gradient is used to elute the adsorbed material.

Immunoblotting (Western blotting) A method using antibodies to label proteins of interest after their transfer onto a membrane support, usually carried out after electrophoretic separation of proteins.

Immunogen The preparation of antigen used to immunize an animal.

Immunohistochemistry Using specific antibodies to localize a given antigen molecule within a section of tissue.

Immunoprecipitation An undefined mixture of proteins is mixed with a specific antibody. The antibody, together with any antigen from the mixture to which it has bound, is separated from the mixture by binding it to immunoglobulin-binding polymer beads (often Staphylococcal Protein A-coated agarose beads). This technique allows the amount of the antigen within the unknown mix to be determined.

Immunosuppression A process in which the normal immunological defenses of the body are inactivated or reduced; it is necessary to induce this state by the means of drugs in some cases where tissue or organ rejection is likely.

Intraocular Placement within one of the eye chambers, a place which sustains growth well and provides easy access for evaluation of successful growth.

Intraventricular Placement within one of the cerebral ventricles.

Intronless gene A gene that lacks the intervening (non-coding) sequences called introns.

Ion activity The effective concentration of an ion in solution. It is different from the total concentration because not all ions are 'free': some are bound to proteins and others form complexes. The activity (a) of an ion in solution is related to its concentration (m) by an 'activity coefficient' γ such that $a = \gamma m$.

Ion exchange A chromatographic technique which separates molecules by their relative abilities to make ionic interactions with either a positively (anion-exchange) or negatively (cation-exchange) stationary phase at a given pH.

Ion-sensitive fluorescent dye A dye whose excitation or emission spectrum changes when it binds a particular ion.

Iontophoresis (microelectrophoresis) A technique by which pharmacologically active molecules (e.g. drugs or neurotransmitters) in solution within a glass micropipette are ejected into brain tissue by passage of an electric current through the micropipette.

Inside-out patch A 'cell-free' patch clamp recording configuration, in which a small patch of cell membrane spanning the tip of a patch electrode has been pulled away from a cell. The former cytoplasmic side of the membrane now faces the bath and its former extracellular side faces the interior of the patch electrode. Opposite of outside-out patch.

Interference filters Coated glass plates of a certain thickness designed to cause all wavelengths, apart from the desired wavelength, to destructively interfere within the filter.

Intracerebral microdialysis Technique for measuring extracellular levels of endogenous (or exogenously applied substances) in specific brain regions of experimental animals. Can be applied to either anaesthetized or conscious animals.

Intron The non-coding part of genomic DNA. Introns are transcribed along with exons to form heterologous RNA (hnRNA).

In vitro Literally 'in glass', i.e. a process carried out in a test tube or in tissue culture.

In vivo In the living animal.

Lamba bacteriophage A specific form of bacteriophage that naturally infects *Escherichia coli* (E.Coli). It can either insert into a specific site along the *E. coli's* chromosome or remain as a virulent phage in the bacteria's cytoplasm.

Leaky-integrator An electrical device comprising a resistor and a capacitor, the values of which determine for how long a charge will be stored in, or leak out of, the capacitor (termed the time constant, τ, where $\tau = RC$). It is used for quantifying a multi-unit EMG signal (τ, usually 100 ms) in terms of amplitude or area of the signal, but has the disadvantage that some of the fine temporal information contained in the original signal is lost.

Lipophilic 'Fat loving'; soluble in fat.

Locus coeruleus Literally 'blue place', the term refers to a pigmented nucleus on the floor of the fourth ventricle of the brain.

Microelectrophoresis (iontophoresis) A technique by which pharmacologically active molecules (e.g. drugs or neurotransmitters) in solution within a glass micropipette are ejected into brain tissue by passage of an electric current through the micropipette.

Microsomes Small membrane vesicles formed after tissue homogenization. They are derived mainly from the endoplasmic reticulum but can also include plasma membranes, endosomes, lysozymes and golgi membranes.

Microtome Machine for cutting thin sections of tissue for microscopic examination.

Mismatch string A string of nucleotides that differs from another strand by only one or two nucleotides.

Mobile phase The liquid (in liquid chromatography) which moves through a chromatographic system.

Monochromator A light source which produces specific wavelengths by illuminating a diffraction grating with white light and focussing the required wavelength into a microscope.

Monopolar electrode An active recording electrode that can placed in close proximity to the biological signal of interest; the reference electrode is placed on 'inert' tissue. A tungsten microelectrode used for human microeneurography is an example of a monopolar electrode.

Motor unit A motoneuron and the muscle fibers it supplies.

Moving average Calculation of a signal mean from a digitized signal by computing the average of a number of data points over, say, 10 points before and 10 points after each data point. The moving average is the resultant mean of each of the averaged values calculated over each sampling epoch; it will have fewer data points and hence less noise. This provides a signal that is more true to the recorded signal (e.g. EMG), in that it does not introduce phase delays.

Myeloma A cancer cell line derived from the B-lymphocyte cell lineage. It has the useful property of being able to continue to divide forever in culture (immortalized).

Neurites Neuronal processes (axons and dendrites), especially of cells in development.

Neutral density filter A filter which attenuate all wavelengths of light to approximately the same degree, so reducing the stimulus intensity.

NMDA *N*-methyl-D-aspartate, an agonist to a specific type of glutamate receptor.

Non-, jumbled-, or scrambled-strings Strings of nucleotides that differ from another string by more than a few nucleotides.

Northern blotting A filter hybridization technique for the identification and localisation of a specific RNA molecule after gel electrophoresis.

Nyquist frequency This is equal to double the highest frequency component of the signal being measured. The Nyquist frequency is the minimum sampling rate which does not produce aliasing of the measured signal.

Nucleotide Purine (adenine and guanine) or pyrimidine (thymidine, uridine and cytosine) bases linked to either a ribose or deoxyribose containing a 5′ phosphate group.

Oligonucleotide Multiple nucleotides linked by phosphodiester bonds between the 3′ hydroxyl group of the sugar of one nucleotide and the 5′ phosphate group of an adjacent nucleotide.

Optimal bilayer A bilayer that has a negligible leak current, high capacitative current and stable electrical properties.

Organotypic slice Term used for a culture in which a thinly-sliced but otherwise intact piece of nervous tissue is maintained in culture medium in an incubator for a day or more (up to several weeks). Organotypic slice cultures can maintain much of their synaptic architecture (cf. cultured dissociated neurons).

Outside-out patch A 'cell-free' patch-clamp recording configuration, in which a small patch of cell membrane spanning the tip of a patch electrode has been pulled away from a cell. The former extracellular side of the membrane now faces the bath and its former cytoplasmic side faces the interior of the patch electrode. Opposite of inside-out patch.

Passage of cells A tissue culture term referring to the process whereby cells are grown on dishes, disrupted and then re-plated onto new dishes (one passage).

Patch-clamp General term for a suite of techniques that depend upon the ability to tightly seal the tip of a patch electrode against a cell membrane. The patch of membrane under the electrode is then electrically and mechanically isolated, permitting low-noise recording of single channels in the patch or, by rupturing the patch, low-resistance access to the interior of the cell.

Patch electrode A glass pipette, filled with salt solution, used for making patch-clamp recordings. Typically, it has a tip diameter of $\approx 1\ \mu m$. The shank of the electrode may be coated with a hydrophobic substance (e.g. Sylgard) to reduce its capacitance, and its tip may be lightly melted ('polished') to smooth out any sharp edges that might damage the cell membrane.

Patch-slice A term for the application of patch-clamp to acute slices of brain tissue. There are two main variants: (i) 'blind' patch-slice, in which a patch electrode is pushed into a slice using only a low-magnification microscope for coarsely positioning the electrode in the region of interest, and (ii) 'visualized' patch-slice, in which a higher-powered microscope with contrast enhancement is used to target selected neurons within the slice.

Percutaneous Literally, 'through the skin'.

Perforated patch-clamp A variant of whole-cell patch-clamp in which a pore-forming agent, such as nystatin, amphotericin or gramicidin, is added to the internal solution in the patch electrode. Rather than rupturing the membrane under the electrode tip, as for normal whole-cell recordings, one simply waits for the pore-former to incorporate into the membrane patch. This allows electrical access to the interior of the cell via the pores, but does not allow 'washout' of larger cytoplasmic molecules.

Peristimulus time histogram This is a common way of representing the temporal (i.e.

Phenotype A measurable characteristic (physical, biochemical or molecular) of a living organism.

Photobleaching The breakdown of dye molecules during exposure to light.

Photoisomerization A very rapid alteration in the shape of the rhodopsin molecule which occurs in response to the capture of a photon. The first step in phototransduction.

Photomultiplier A light detector based on an analog valve. It produces a voltage signal proportional to the amount of light illuminating the front aperture.

Photoreceptors Rods and cones; the neural elements of the retina which capture photons and generate an electrical signal that is transmitted to the retinal bipolar cells.

Phototransduction A biochemical cascade of events within the photoreceptor outer segment beginning with the photoisomerization of rhodopsin, activation and amplification through a G-protein, activation of the cyclic GMP-phosphodiesterase enzyme, hydrolysis of cyclic GMP, and closure of sodium channels in the outer plasma membrane with consequent hyperpolarization of this structure.

Piezoelectric A substance which changes shape in response to an applied voltage.

Piezoelectric bimorph A translator built from a two piezoelectric wafers glued together and wired in opposing directions. It produces movement when the wafers bend in response to an applied voltage. It is relatively slow, but can produce large movements (>1 mm).

Piezoelectric stack A translator built from a stack of piezoelectric wafers. It produces movement when the wafers expand along the length of the stack. It is very fast, but only produces small movements (<100 μm).

Piezoelectric translator A device based on a piezoelectric substance that moves a load in response to a changing voltage.

Pigment epithelium Layer of cells containing pigment granules behind the layer of photoreceptors in the eye.

Plasmid A double-stranded circular DNA that can replicate in bacteria. It is used for the majority of routine DNA manipulations.

Pneumothorax Presence of air in the pleural cavity around the lungs; normally the pleural cavity is a potential space with no air.

Poles (filter) This describes the rate of attenuation of the filter frequency response above the corner frequency. The term stems from the mathematics used to describe the filter response. As a working definition an N-pole filter has the same role as N single pole filters (with different corner frequency) connected in series.

Polyclonal antibodies A mixture of antibodies hopefully all directed against a common antigen molecule but often recognising different epitopes and with differing affinities.

Polyclonal antibody preparation A variety of antibodies to different parts the same molecule.

Polymerase chain reaction A technique that amplifies specific nucleotide sequences *in vitro*.

Primary culture A culture in which the cell of interest no longer undergoes cell division. Thus, the cells steadily die off with time, and new cultures must be periodically established by taking fresh tissue from an animal. Most neural cultures are primary cultures. See also cell line.

established by taking fresh tissue from an animal. Most neural cultures are primary cultures. See also cell line.

Primary ion The primary ion of an ion-selective electrode is the ion to which the electrode has a selective response.

Primer A short strand of DNA (oligonucleotide) that is complementary to a region on the template DNA or mRNA. DNA synthesizing enzymes cannot operate without its presence.

Proliferation The process by which cells grow to larger numbers of the same type of cell.

Promoter A nucleotide sequence in DNA to which RNA polymerase binds to begin transcription.

Pronucleus The nuclei of the gametes (sperm or ova) each containing 1 set of chromosomes. Upon fertilization of the female egg (ova) by the sperm, the two pronuclei exist within the cytoplasm of the egg for some six hours. They then combine to make a diploid cell.

Protease An enzyme which hydrolyses peptide bonds and thus degrades proteins.

Protein A-sepharose Protein A is a bacterial protein that binds to certain antibodies. Sepharose is a non-biological substance usually made as small beads and is produced by a company called Pharmacia. Molecules like protein A can be attached, resulting in protein A-sepharose. This can then be used to bind and purify antibodies from serum because the sepharose is easily recovered from a solution.

Pseudogenes Genes that have a similar sequences to the functional genes but have been disabled by mutations that prevent their expression.

Radiolabelling Process of applying a radioactive tag to a cellular constituent.

Ratioing Technique of comparing fluorescence at two different wavelengths to obtain a measure of ion concentration.

Recombinant DNA A hybrid DNA molecule in which DNA from two organisms has been spliced together using the techniques of genetic engineering. (Note: recombinant DNA also occurs naturally as a result of crossing over during recombination but common usage of the term implies genetic engineering.)

Rectification A signal-processing technique in which, for example, the electrical sign of negative-going parts of a signal are changed to positive and then added to the original positive-going parts of the signal. This is used for averaging the EMG records of several trials because otherwise the positive and negative components of the signal would tend to cancel each other out.

Reporter gene Any introduced gene whose product can be detected in tissues. Ideally the gene should not normally be present in the cell that is to be transfected with the reporter gene. In mammalian studies bacterial genes such as *LacZ* which encodes for β-galactosidase enzyme is used as a reporter gene.

Retaining current The current used during microelectrophoresis to prevent diffusion of a molecule from the tip of the micropipette.

Retrograde Travel in a backward direction, i.e. towards the neuronal cell body.

Retrovirus A virus in which the genetic material is RNA and the enzyme reverse transcriptase is used to produce a complementary DNA molecule in the host cell.

Reversed phase chromatography A chromatographic technique which is used to separate molecules on the basis of their relative hydrophobicity. Organic solvent is used to elute the adsorbed material. The mobile phase is more polar than the stationary phase, in contrast with 'normal' phase chromatography where the opposite is true.

Reverse transcriptase An enzyme of retroviruses which synthesizes a DNA chain on an RNA template.

Rhodopsin The photopigment within the rod photoreceptor outer segments which captures light.

Ribonuclease An enzyme that degrades RNA and is highly resistant to denaturation.

Ribonucleic acid (RNA) Bases that are transcribed from the gene to form messenger RNA (mRNA).

Riboprobe *In vitro* synthesized linear strand of ribonucleic acid labelled with either a radioactive or non-radioactive tag and used in hybridization reactions.

RMS processing Calculation of a signal mean by computing the root-mean-square (RMS), a mathematical measurement of signal power. This is advantageous for EMG recordings because, after rectifiying the signal and performing RMS processing, the overall shape of the signal closely reflects the variations in intensity of the original signal.

RNA polymerase The enzyme which catalyses the synthesis of an RNA molecule from a strand of DNA.

Sclera Fibrous outer layer of the back of the eye.

Scotopic Low levels of light. Rod photoreceptors function optimally in scotopic conditions. The electroretinogram (ERG) elicited from the dark-adapted eye by light stimuli of low intensity is referred to as a scotopic ERG—the rod contribution to this ERG is maximized, whereas the cone contribution is minimized. The converse situation is referred to as the photopic ERG.

Selectivity (of an ion-selective electrode) A description of the response to the primary ion in relation to the response to any interfering ion. A highly selective electrode will show only a response to the primary ion and changes in the concentration of other ions will not affect the voltage recorded.

Sense string A string of nucleotides that codes genetic information.

Sensitivity (of an ion-selective electrode) The change in voltage recorded by the electrode due to a change in concentration of the primary ion. An ideally sensitive electrode will behave in a Nernstian fashion, i.e. if the primary ion is monovalent there will be $\approx$60 mV change in voltage recorded for a 10-fold change in ion concentration.

Serum The acellular component of blood.

Sham A control experiment where all the manipulations or surgeries are performed but no damage is inflicted.

Sharp electrode A very fine-tipped (orifice usually less than 1 μm) glass microelectrode used to penetrate a cell membrane, in order to measure transmembrane potential or current.

Silanization (of an ion-selective electrode) Under normal conditions the surface of glass is hydrophilic, whereas the liquid containing the ionophore which forms the ion-selective membrane is hydrophobic. To introduce this liquid into the tip of the electrode the glass surface must also be rendered hydrophobic. This is done by a process called silanization where a silane is introduced into the pipette, hydroxyl groups on the surface of the glass react with the silane and are replaced by hydrophobic groups.

Single-cell PCR A method for using a patch electrode to apply PCR ('polymerase chain reaction') to the messenger RNA found within a single cell.

SIT camera Silicon Intensified Tube camera, with an image intensifier mounted in front of a standard video camera.

Solenoid valve A valve which turns a flow of solution on or off in response to an applied current pulse.

Southern blotting A filter hybridization technique for the identification and localization of a specific DNA molecule after gel electrophoresis.

S-phase The phase in the cell cycle that is concerned with DNA synthesis. The cell cycle is a series of steps that begins with mitosis (M-phase), followed by the G1 phase (a gap of time just after mitosis), then the S-phase (DNA synthesis), and ends with the G2 (post-mitotic phase).

Spikes Another name for action potentials; so called because of the spike-like nature of the potentials.

Splice variants Different strands of mRNA formed from hnRNA (heterologous RNA) by differential splicing (see Intron).

Staphylococcal protein A A bacterial protein that binds Immunoglobulin Gs.

Stationary phase In column chromatography, the stationary column packing through which the liquid (mobile) phase traverses.

Stereotaxis Precise localization in three-dimensional space; often refers to locations within the cranial cavity with respect to certain brain structures.

Stochastic Something which occurs as a result of a random process.

Stokes shift The difference, in nm, between the excitation and emission wavelengths of a certain dye.

Stringency Conditions of hybridization which determine whether denatured nucleic strands renature specifically. At high stringency only complementary strands will hybridize. Stringency is determined by salt and formamide concentration, and incubation temperature.

Substantia nigra Literally 'black substance', the term refers to a portion of the midbrain involved in movement; cells in this area deteriorate during the onset of Parkinson's disease.

Sutures Furrows at the junction of adjacent bones of the skull.

Synaptosome Resealed pre-synaptic nerve terminal formed after homogenization of brain tissue.

Time constant (of an electrode) An electrode has resistive and capacitive properties in the same way that a cell membrane does. These properties mean that when current is passed through the electrode there will be a period of time before a steady-state voltage is reached. The initial portion of the voltage change per unit time follows an exponential relationship and the time constant is the time taken to reach $\approx$0.63 of the steady-state voltage change.

Transcription The copying of one strand of DNA into a complementary RNA sequence by an enzyme called RNA polymerase.

Transfection The update of exogenous DNA by mammalian cells in culture, most typically achieved by forming a calcium phosphate/DNA precipitate which is applied to the cells.

Translation The process by which the nucleotide sequence in mRNA is used to incorporate the appropriate amino acids into protein.

Transplantation Transferring part of the body from a donor to a host; in non-clinical settings, often referred to as grafting.

Tritium Radioactive isotope of hydrogen with an atomic weight of three.

Trophic factors Substances which foster growth; contrast with *tropic*, which indicates a directional attraction of growth.

Tropic Having a directional influence on growth.

Vacuum transfer Application of a vacuum to enhance the transfer of nucleic acids from a gel to a nitrocellulose filter or nylon membrane.

Vector Often used to refer to some means (e.g. a recombinant retrovirus) for introducing, and expressing, a cloned gene in a host cell.

Vehicle Inert medium in which a biologically active agent is administered.

Vibratome Machine which uses a vibrating knife to cut tissue sections for microscopic examination.

Volley A barrage of neural impulses, e.g. a sensory volley is composed of action potentials propagating up a nerve and originating from many sensory receptors.

Voltage-clamp A technique used to measure transmembrane currents while the transmembrane potential is held constant at a chosen value.

Watson–Crick base pairing Bases on opposite strands of DNA are bound by hydrogen bonds. Adenine specifically binds to thymidine and cytosine binds specifically to guanine. This pairing is referred to as base-pair complementarity.

Whole-cell recording A patch-clamp recording configuration in which a patch electrode has been sealed against the surface of a cell and the small patch of membrane under the electrode tip has been ruptured, usually by a pulse of suction. This makes the interior of the patch electrode contiguous with the interior of the cell, permitting good electrical access and solution exchange between the electrode and the cell's cytoplasm.

Xenograft Tissue or organs taken from one individual and transferred to another individual of a different species; synonymous with heterograft.

***Xenopus* oocyte** The unfertilized egg of the African clawed toad, about 1 mm in diameter, into which RNA can be injected for translation and cell surface expression of ion channels.

INDEX

Page numbers in bold indicate a definition either in the glossary or within the text.